THE STEPHEN BECHTEL FUND

IMPRINT IN ECOLOGY AND THE ENVIRONMENT

The Stephen Bechtel Fund has

established this imprint to promote

understanding and conservation of

our natural environment.

The publisher gratefully acknowledges the generous support to this book provided by the Stephen Bechtel Fund.

Serendipity

ORGANISMS AND ENVIRONMENTS
Harry W. Greene, Consulting Editor

Serendipity

AN ECOLOGIST'S QUEST
TO UNDERSTAND NATURE

James A. Estes

UNIVERSITY OF CALIFORNIA PRESS

University of California Press, one of the most distinguished university presses in the United States, enriches lives around the world by advancing scholarship in the humanities, social sciences, and natural sciences. Its activities are supported by the UC Press Foundation and by philanthropic contributions from individuals and institutions. For more information, visit www.ucpress.edu.

University of California Press
Oakland, California

Library of Congress Cataloging-in-Publication Data

Names: Estes, J. A. (James A.), 1945– author. Title: Serendipity : an ecologist's quest to understand nature / James A. Estes
    Description: Oakland, California : University of California Press, [2016] | 2016 | Includes bibliographical references and index.
    Identifiers: LCCN 2015048153 | ISBN 978-0-520-28503-3 (hardcover : alk. paper)
    Subjects: LCSH: Sea otter—Ecology—Alaska. | Marine ecology—Alaska. | Biotic communities—Alaska.
    Classification: LCC QL737.C25 E85 2016 | DDC 599.769/509798—dc23
    LC record available at http://lccn.loc.gov/2015048153

Manufactured in the United States of America

25 24 23 22 21 20 19 18 17 16
10 9 8 7 6 5 4 3 2 1

In keeping with a commitment to support environmentally responsible and sustainable printing practices, UC Press has printed this book on Natures Natural, a fiber that contains 30% post-consumer waste and meets the minimum requirements of ANSI/NISO Z39.48–1992 (R 1997) (*Permanence of Paper*).

To Dorothy, for teaching me discipline,
and to Frank, for teaching me how to dream

CONTENTS

FOREWORD

Serendipity: An Ecologist's Quest to Understand Nature is the fourteenth installment in the University of California Press's Organisms and Environments series, whose unifying themes are the diversity of life, the ways that living things interact with each other and their surroundings, and the implications of those relationships for science and society. We seek works that promote unusual, even unexpected, connections among seemingly disparate subjects, and that are distinguished by the unique talents and perspectives of their authors. Previous volumes have spanned topics as diverse as bison natural history and adaptive radiation in lizards, but none have addressed ecological relationships with the detail explored herein.

Serendipity recounts James Estes's illustrious, career-long focus on a region and its organisms, some of them familiar and charismatic, others less so, with an emphasis on untangling ecological processes. Let's be clear: Though 'ecology' is commonly used to describe the state of our planet or indicate a love for nature, scientists who study ecology are seeking to comprehend relationships among organisms and their surroundings, including living and non-living components. It was originally spelled 'oecology,' its root being the Greek *oikos,* for "house or home." Typically our discipline is conceptualized in terms of population biology, focused on births, deaths, and how they affect overall numbers of *a particular species;* community ecology, concerned with predator-prey dynamics, mutualisms, and other interactions *among species;* and ecosystem studies, which explore the significance of energy flow, nutrient cycling, and other *abiotic factors* for populations and communities. Our understanding of ecological relationships—population, community, and ecosystem processes—inevitably plays essential roles in conservation, including an appreciation for particular places and their inhabitants.

Jim's story begins in the Aleutians, volcano-studded islands that stretch scythe-like from the Alaskan Peninsula toward Russia between the Pacific Ocean and Bering Sea. As his research unfolded, it extended up into the Arctic and thence south to California. Later his research embraced more global perspectives. At a finer scale, *Serendipity* proceeds along rocky shorelines characterized by frigid tides and stunted coastal vegetation, places so harsh that biologists wear flotation suits for protection in case of an accidental dunking. Those simple Aleutian ecosystems are repeated dozens of times, the islands differing mainly in terms of human impact, and they thus proved especially useful for answering community-level questions. Initially Jim and his colleagues studied sea otters, the largest members of the weasel family; sea urchins, saucer-size herbivorous relatives of sea stars; and kelp, a brown alga whose stem-like stipes and leaf-like blades can grow more than 200 feet long. Jim's graduate research confirmed that otters suppress urchin numbers, thereby preventing overgrazing and permitting dense underwater algal forests, but that absent predation, urchin populations explode and kelp crashes—and that's just for starters!

This also is a memoir of how science often transpires, pursued as it is by people whose lives are influenced by unexpected circumstances and their need for adventure—Jim avoided the Vietnam-era draft thanks to a popping knee, and shortly thereafter was off to the Aleutians for reasons having little to do with community ecology. No doubt Pasteur's aphorism, that chance favors a prepared mind, soon came into play for him, too, as well as boundless curiosity and a powerful work ethic. Now, thanks to more than 40 years of measuring, counting, analyzing, getting grants, and writing up results, the otter–urchin–kelp linkage is a textbook example of so-called trophic cascades, and Jim has been honored by election to the National Academy of Sciences. Moreover, humans and whales have recently synergistically altered his study systems, prompting still more surprising questions as well as an unanticipated collision of science, personalities, and politics.

Serendipity: An Ecologist's Quest to Understand Nature provides an accessible, scholarly account of a major scientific advance, one that's still unfolding with humans' ever-growing environmental impact. For seasoned professionals, *Serendipity* synthesizes a brilliant career, yielding far more than the sum of individual research projects; for the next generation of field biologists, it's a look at how it all gets done and some hints to the path ahead. Lay readers as well will enjoy Jim's book for its lucid explanations of quantitative research, combined with his rich vision of what's going on in marine intertidal communities. And perhaps most profoundly, *Serendipity* carries a lesson for everyone concerned about the fate of biological diversity: Conservation is not just about fascinating sea otters and peculiar urchins, stately kelp and awe-inspiring killer whales—it should also be about preserving their interactions. At the same time, those component species themselves matter, especially when, as is the case with apex predators, they play central roles in structuring ecosystems.

Harry W. Greene

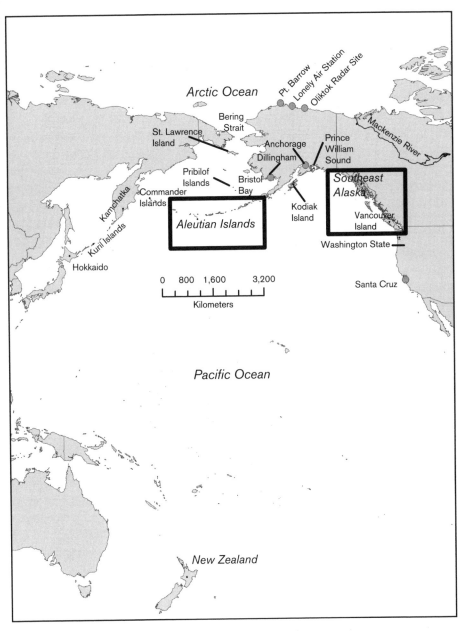

MAP I. The Pacific basin.

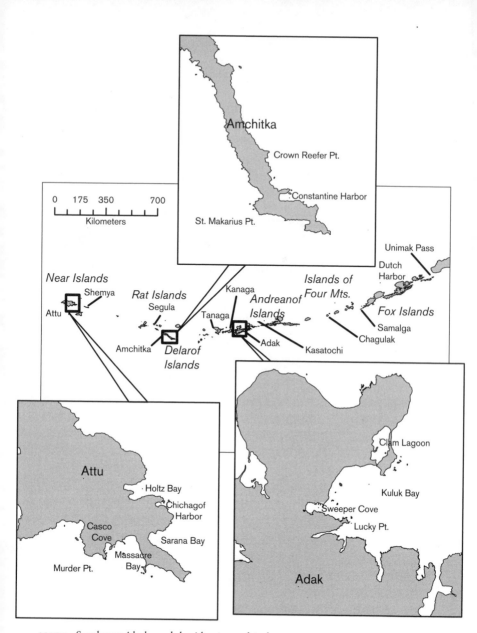

MAP 2. Southwest Alaska and the Aleutian archipelago.

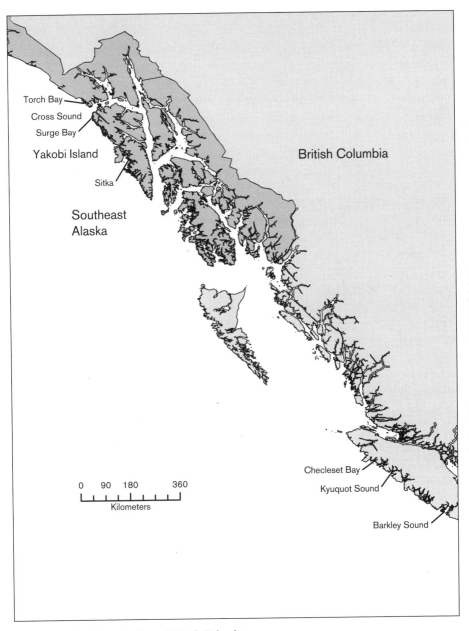

Torch Bay
Cross Sound
Surge Bay
Yakobi Island
Sitka
Southeast
Alaska

British Columbia

Checleset Bay
Kyuquot Sound
Barkley Sound

0   90   180   360

Kilometers

MAP 3. Southeast Alaska and British Columbia.

In the Beginning

IT WAS MAY 1970. I had finished my master's degree the previous spring, but the Vietnam War was raging and the draft board was hot on my heels. Military service seemed imminent, and my thoughts were mostly on how to do this as quickly and safely as possible. I had been weighing two choices: the draft and a two-year stint in the army (short but dangerous, and with little imagined value) versus officer training and four years in the navy (longer but safer, and more interesting). Four years seemed like a long time, so I opted to take my chances with the draft.

I was single, fit, and in excellent health—seemingly an ideal prospect for military service. After receiving my notice to report for the preinduction physical, I showed up at the Spokane induction center fully expecting this to be the first step toward boot camp and a stint in the army. Draft dodging was in vogue in those days, and many of my fellow preinductees had letters from their physicians, psychiatrists, preachers, and the like, explaining why they were physically or psychologically unfit to serve. The induction staff had seen it all before and regarded every one of these would-be draft dodgers with obvious disdain. There were hundreds of us, assembled alphabetically in lines and ordered to strip for what amounted to little more than confirmation that major body parts were intact. Everyone passed. This was followed by intelligence and mental-stability tests. To my amazement, a few of the preinductees actually managed to fail this second part of the exam. But the saying in those days was "If you have a pulse, you're fit to serve," and I had little hope of not making it through the perfunctory battery of examinations.

The last step of the preinduction physical was a short interview with a military physician. It was at this point that the paper waving and ululating reached their crescendos. My examining physician, a tough-looking marine named Sanchez, patiently listened to each man's story, looked over his paperwork, and then stamped him "fit to serve." When my turn came, Dr. Sanchez seemed surprised by the lack of excuses and said as much. At the end of the interview, he asked if I was sure there wasn't something physically or mentally wrong with me that should be noted in the paperwork. I thought about it more carefully and mentioned a shoulder injury that had ended my career as a baseball pitcher but otherwise had never bothered me. Dr. Sanchez seemed unimpressed, but the discussion prompted him to ask of any other orthopedic problems. Since childhood, my right knee has popped when I move my leg in just the right way, but the popping knee had never bothered me (and now, at 70, still hasn't). It must have sounded scarier than it was, because the doctor's eyes widened when he heard the snap, and in that instant I was deemed physically unfit to serve, classified 4-F. I could barely contain my euphoria, and after gathering my belongings, I headed for the nearest bar to celebrate my freedom. Thank you, Dr. Sanchez.

Suddenly, I was forced to think seriously about what to do with a newfound life. I wanted adventure and I wanted to continue my education in biology. While a graduate student at Washington State University, I had developed a friendship with Professor Vincent Schultz, a statistical ecologist and consultant to the Atomic Energy Commission (AEC). The Cold War was in full force, and the AEC was actively engaged in underground nuclear testing. Most of these tests were done at the Nevada test site, about 65 miles from Las Vegas. But the AEC wanted to test a much larger device, and they needed a more remote place to do that. The chosen site was Amchitka Island in the western Aleutian archipelago.

Amchitka was about as far from human habitation as one can get within the U.S. territorial boundary. It also had a fine deep-water harbor and a long and still functional runway built in World War II, during staging for the invasion of Japan. But the proposed test was deeply unpopular, especially in Alaska, where the AEC's credibility had been strained by Operation Chariot (O'Neill, 2007). The perils of radiation were well known and there was concern that the detonation of such a powerful blast might vent to the atmosphere or leak into the surrounding ocean, thus contaminating the rich North Pacific fishery and posing a health risk to people in Alaska as the imagined radioactive cloud drifted eastward with prevailing winds. There were also concerns over the impact of such a powerful detonation on the environment of Amchitka itself, which was part of the Alaska Maritime National Wildlife Refuge and home to numerous species, including an abundant sea otter population. The welfare of these animals had become a point of particular concern and controversy. The AEC asked Vince Schultz to help them find someone who could measure these impacts, and in June he called to ask if I wanted the job. By early October I was off to the Aleutians, intent on spending the next two years with sea otters at Amchitka Island.

I never imagined at the time that this early experience would lead to a life of such opportunity and adventure. I've returned to the Aleutian Islands nearly every year since, fascinated by the ecology and natural history of the place and following one question after another as my understanding of ecological concepts and the natural history of the area grew. Almost 45 years later, as I sit down to write this book, my goal is twofold. First and foremost, I wish to recount the science—the patterns of nature that I have endeavored to understand, what I have learned, and how I learned those things. Most of this has been published in the scientific literature, and some of it has been recounted in various textbooks and other secondary sources. This book is not a rehash but rather an integrated overview built around my own history and the history of the region. My personal history is important because it defines a chronology of discovery and understanding as one step slowly and often unexpectedly led to the next. The region's history is important because it affords a window for understanding the science. My second goal is to explain how the science really happened. Although I had what I thought was a clear and exciting vision from early on, the truth is that my vision changed markedly along the way, informed by fortuitous meetings with a few key individuals and the serendipitous influence of events that could never have been foreseen or understood as relevant at the time. Every scientist is deeply influenced by these latter kinds of experiences, but few ever write about them. As a result, the lay public and even nascent scientists have distorted views of the way interesting science is commonly done.

This book uses sea otters and *kelp* forests as examples to demonstrate ecological processes, in particular how species are linked together in *food webs* and the dynamic properties of the linkages. Tracing these patterns is relatively easy, requiring only the ability to identify species and the patience to watch, or otherwise figure out, who is eaten by whom. Understanding the dynamic properties of these linkages (the process of nature) is more complex and much more challenging. For example, for each particular consumer–prey linkage, we would like to know (a) to what extent the distribution and abundance of the *consumer* are influenced by the *prey* and (b) to what extent the distribution and abundance of the prey are influenced by the consumer. We'd like to know how these direct linkages of consumer and prey are themselves linked in *food chains* and—given the nearly infinite number of potential such pathways in the resulting food web—which of those that actually exist really matter. And we'd like to know how these systems were assembled over the long sweep of time, the reciprocity between ecology and evolution that played out during that assembly, and the degree to which ecosystems are being disassembled or rearranged by the pervasive hand of human influence.

The challenge in understanding dynamic process is that these forces of nature are invisible to the glance, and can remain unseen no matter how intently one looks. Only by experimentally perturbing the distribution and abundance of species can we see and understand the linkages between them. And therein lies the magic of the Aleutian Islands and their surrounding kelp forests. Unlike tropical rainforests or coral reefs, these kelp forests contain relatively few species. In contrast to most tropical islands, the topography and prevailing winds of the Aleutians have little or no influence on biotic communities, adding further to the biological simplicity of these islands. Also in contrast to tropical islands, the ubiquitous scouring influence of *Pleistocene* glaciers has largely eliminated regional variation from the evolution of island endemic species. What we have instead are hundreds of intrinsically similar islands, differing from one another largely as a consequence of the things people have done to them. These are the *perturbations* that make the Aleutians such a marvelous place for learning about the dynamics of nature.

My particular object of interest has been the sea otter and how this predator influences the structure and organization of food webs in kelp forests. Sea otters once occurred in abundance across the North Pacific rim but were hunted almost to extinction during the Pacific maritime fur trade. By 1911, when sea otters were finally protected from further takes, only two or three of the Aleutian Islands supported small remnant populations. They were extinct elsewhere. With protection, the remnant populations began to recover and spread. Almost 60 years later, when I arrived on the scene, sea otter populations had recovered to historical levels at some islands, were at various stages of recovery at others, and remained absent from still others. This is the perturbation I have used to explore the sea otter's influence on kelp forest food webs. In the beginning I simply compared rocky reef communities between islands with and without sea otters. Later my colleagues and I expanded that approach to southeast Alaska and British Columbia. In the mid-1970s I began watching systems change as they were reinvaded by sea otters when the populations of the latter increased and spread. Then, in the mid-1990s, I watched these changes run in reverse as sea otter populations suddenly collapsed, apparently after they were discovered and gobbled up by killer whales.

Much earlier in life, as I looked forward to a career in science, the prescription for success seemed vague and daunting. Looking back now, it all seems clear and simple. Thus, the book is structured as an account of how my career unfolded and the things I've learned along the way. Chapter 2 establishes the intellectual context of the story. Although sea otters and kelp forests provide the particulars, my work is really about large *apex predators* and their influences on ecosystems. I explain in detail the conceptual issues, the challenges to understanding these issues, the approaches scientists have taken in an effort to gain understanding, and why it is important. In chapter 3, I move to a specific focus on the natural and human history of the Aleutian archipelago and how the essentials of these histories have made that place so interesting. Chapter 4 recounts my initial discovery of the sea otter's role in kelp forest community structure, beginning with my earlier perceptions after spending nearly a year on Amchitka Island and the exciting "aha moment" that came with my first short trip to Shemya Island the following summer.

After completing my graduate studies on sea otters and kelp forests, I was hired by the U.S. Fish and Wildlife Service to conduct research on arctic marine wildlife. My first assignment was to survey the abundance and distribution of Pacific walruses, with the understanding that further work on walruses and their arctic marine ecosystem would follow. Chapter 5 provides an overview of what I did, what I learned from it, and why those experiences led me back to sea otters and kelp forests. In chapter 6, I begin to lay out the questions and challenges that followed my earlier discovery of the sea otter's ecological role as a *keystone species* in kelp forest ecosystems. My original studies were done in one small corner of the North Pacific, thus begging the question of whether the same ecological processes occurred elsewhere. Chapter 7 is focused on efforts to assess generality and variation in those processes across the North Pacific Ocean and southern Bering Sea, and to understand how and why these systems change with changes in sea otter population density.

Like forests everywhere, kelp forests provide a home and important resources to numerous other species. In chapter 8, I explain the key processes involved in these associations and how they influence the behavior and population biology of certain species. The material covered through chapter 8 establishes a number of strong linkages, between species or between *functional groups,* that constitute ecological forces with expected evolutionary consequences if played out over large enough areas and sufficient lengths of time. In chapter 9, I recount what we know or suspect about the evolutionary consequences of sea otter predation. Although these explorations might have gone in any number of directions, my work was focused on how variation in food-chain length has influenced the *coevolution* of defense and resistance in plants and their associated herbivores, how these apparent effects have influenced the ecology and evolution of other kelp consumers, and how the coevolution of plants and their herbivores has influenced both the structure and dynamics of the kelp forest ecosystem.

My story to this point was scientifically exciting but politically benign. That all begins to change in chapter 10 with the unexpected collapse of sea otter populations in the Aleutians and the conclusion that killer whale predation was the likely cause. I will explain in detail what I saw (the beginnings of the sea otter collapse in southwest Alaska), what I did retrospectively in an effort to understand the cause, and why I concluded that the likely culprits were killer whales. Although this conclusion seemed to resonate well enough with most others in the scientific and management communities, it left the question of "why" unanswered. In chapter 11, I recount the data, analyses, and logic that led my colleagues and me to recognize that the declines of sea otter populations were only part of a larger collapse of other coastal-living marine mammals in the same region, and to conclude that post–World War II industrial whaling was the likely cause. That conclusion, to put it mildly, didn't resonate so well with others who had been working on these other species, which leads to chapter 12, "Whale Wars." This was my first exposure to the darker corners of science, and the experience was both intellectually shocking and emotionally devastating. Recognizing that many beliefs are difficult to understand by anyone but their beholders, I'll nonetheless recount some of the key events and my own interpretations of why this work became so controversial.

Chapter 13 moves back to science with an account of the seemingly unrelated interplay among introduced foxes, seabirds, and terrestrial plant communities in the Aleutian archipelago. The foxes were introduced as a substitute fur resource after sea otters had been hunted almost to extinction, so the fox story is legitimately part of the sea otter's ecological legacy. Chapter 14 looks beyond coastal ecosystems of the North Pacific to ask the big question: How much of what we know or believe about that system is unique to sea otters and kelp forests, and how much might apply to other species of apex predators and their ecosystems around the world? Chapter 15 is a retrospective on the science, focusing on approach, scale, and the larger conceptual issues that emerged from my years of work on the sea otter–kelp forest system. Chapter 16 ends the book with a look to the future. What are the prospects for sea otters and kelp forests; what more do we need to know about them and how might the next generation of scientists go about learning those things; and why should that matter to more than a handful of people?

My life as a naturalist and ecologist has taken me to numerous places across the Pacific basin, many of which are remote and poorly known. Familiarity with this geography is of more than passing interest, because most of my work is founded on contrasts between different places or on the patterns of change at particular places through time. The three maps that precede this chapter identify the locations of all place names mentioned in the book.

Part of my job as a scientist has been to write technical reports, papers, and books for a relatively narrow and well-defined audience of other professionals. This book is aimed in part at that same group of professionals, to give them an integrated overview of what I've learned through the years about sea otters and kelp forests. My approach is not to dwell on the technical details but to recount the concepts, ideas, and high points of discovery that tie nearly half a century of study together as an integrated whole. Within this circle of professional ecologists, my particular target audience is students and younger scientists. They are the ones who will carry ecology forward, and I want them to understand how my own contribution to that endeavor actually happened.

I hope to reach a wider segment of society as well. I've tried to make the writing accessible to anyone with an interest in nature. Important terms are italicized when introduced and are defined for easy reference in the glossary, an aid for quickly comparing related concepts. And a complete bibliography gathers 50 years and beyond of ecology for those who want to dive deeper.

Finally, I have written this book for my two children, Colin and Anna, who endured my long absences while puzzling over exactly what I was doing and why it consumed me. Colin and Anna—I know you love and respect me. Now I hope you will better understand me.

TWO

Understanding Nature

MY FATHER WAS A NAVIGATOR who flew supply missions across the Pacific during World War II. A distinct childhood memory from his many stories is of the wretched weather in the Aleutian Islands. This recollection was rekindled in October 1970 as I flew from Anchorage to Amchitka Island for the first time. The air was clear at 30,000 feet, the sky above a brilliant blue, but all I saw below was the gray of clouds and fog. As we descended to land at Amchitka, what first met my eyes was a bleak, treeless landscape surrounded by a sea of crashing waves. The reality of working in such a place was exhilarating on one hand but disconcerting on the other. I wondered for a moment if I might not have been safer in Vietnam. In addition to that disturbing thought was a realization that I had almost no idea what to do. My contract with the AEC only specified that I would conduct such studies as were necessary to document the impacts of their activities on sea otters. Although I'd never even seen a sea otter, I knew I could fulfill my contractual obligation. But I wanted to do more than that—I wanted to use this wonderful opportunity to learn something new and interesting about these animals and the place where they lived. At the time, I had little idea of exactly what that might be.

As I pondered the challenge, my mind was drawn to intriguing features of the otters themselves—that they were the smallest of all marine mammals; that they depended entirely on fur for insulation from heat loss to the cold water in which they were constantly immersed; that they were adept at using rock tools to break through the tough exoskeletons of their invertebrate prey; and much more. I thought about the interplay between these animals and their environment, but only from the perspective of how they met the challenge of providing for their bodily needs, like staying warm, getting enough to eat, and avoiding being pummeled by the often violent surrounding ocean.

I wasn't thinking about their connection to kelp forests, except that these forests provided a place for the otters to live. Beyond that, I had no view of how the sea otter's world was put together and worked. I realized that this too was an important question, but at the time it seemed obtuse and only marginally relevant to what I imagined could be learned from this animal. I thought of ecosystems as the cradle of life, and of predators like sea otters as little more than *ecological passengers,* drawing their various bodily needs from the ecosystem but giving little back in return. That was my worldview of ecology in 1970. Now I see it differently. I still think of ecosystems as the cradle of life, but I understand that species interactions add richness and complexity to nature, and I have come to believe that understanding why ecosystems look and behave as they do is modern science's most pressing challenge.

My view of ecology grew from two early observations. One was the seemingly poor track record of natural resource management, which I think of as the human endeavor of intervening with nature to achieve some desired result. Part of the difficulty is that resource managers must contend with humankind's ever increasing global footprint while never being given the option of true mitigation. But a significant part of the difficulty is that they depend, in their decision making, on scientific algorithms that are either overly simplistic or simply wrong. A second observation that attests to the shallowness of understanding about how ecosystems work is how very often ecologists and natural resource managers are surprised by nature. Disease ecologists rarely see epidemics coming, especially the first time they occur. Invasive-species biologists rarely see new invasions coming, even though they've happened countless times before. Fisheries biologists seldom see particular stock crashes coming, even though there have been numerous such crashes in the past. The list could go on and on. Examples of ecological surprises are everywhere (Doak et al., 2008). All of this is not to say that ecologists know little or nothing of how nature works. In fact they know a great deal. The point is that this vast reservoir of ecological knowledge is only a small and superficial part of the understanding needed to predict the behavior of populations, communities, and ecosystems with any degree of confidence and regularity.

Why is it that ecologists know so little about the workings of nature? Almost everyone knows the answer to this question at one level: simply put, nature is complex. The modern world supports more than a million described species, with an estimated 10 million or more remaining undescribed. And this diversity of species pales in contrast with the diversity of ways in which species can interact with one another (see box 1).

BOX 1. NUMERICAL COMPLEXITY: A LESSON
FOR ASPIRING ECOLOGISTS

I've managed to express most of the ideas in this book without resorting to heavy math, and I'll only spring a few equations on you here. If nothing else, they help convey the complexity I'm talking about.

The challenge in understanding the workings of living nature stems in part from the immense diversity of species and the even greater diversity of ways in which these species can interact with one another. To appreciate this challenge, let's begin by considering just two species living together in the same place. These species might not interact with each other, in which case a change in the abundance of either member of the species pair has no effect on the abundance of the other species. We can characterize this particular situation as 0/0, meaning that neither species has a direct influence on the other, no matter how rare or abundant either might be. But if the two species interact with each other, the possibilities become more complicated and much more interesting. The two species might compete for a common limiting resource, in which case we can characterize the interaction as −/−, meaning that both species incur costs from living together. One species might eat the other species. In this particular case, we can characterize the interaction as +/−, meaning that the consumer realizes a benefit from the interaction whereas the prey incurs a cost. The two species might also realize mutual benefits from living together (+/+), such as occurs between flowering plants and their pollinators; these kinds of interactions are called *mutualism*. Beyond this, the two species might interact with one another in such a way that one member of the pair either realizes a benefit or incurs a cost while the other is unaffected—respectively termed *commensalism* (+/0) and *amensalism* (−/0).

Knowing the qualitative nature of species interactions is essential to understanding ecosystems. But that's only the beginning of what one needs to know. The next important step is to know the strength of the various species interactions. The effect of species 1 on species 2 may be large, small, or somewhere in between. The magnitude of this effect is known as the *interaction strength,* commonly defined as the abundance of species 2 when species 1 is absent minus the abundance of species 2 when species 1 is present (the population-level interaction strength) or by dividing the above value by the abundance of species 1 (the per capita interaction strength).

Interaction strength is an all-or-none measure because it is based on the driver (species 1) being present at some equilibrium abundance. Another important goal in ecology is thus to know the functional relationship in interaction strength between two species as they vary in abundance. In the simplest case, this functional relationship is linear, which is to say that the per capita effect of each of the interacting species on the other species is constant. We make this assumption when we teach introductory ecology students how to model the population dynamics of two interacting competitors or a predator and its prey with simple mathematical constructs like the *Lotka-Volterra equations.* But the functional relationships between two interacting species are frequently nonlinear, which makes both detection and modeling of species interactions trickier.

In a nonlinear scenario, the effect of species 1 on species 2 may be nonexistant or very small until species 1 reaches some critical abundance, at which point species 2 responds sharply, resulting in what ecologists refer to as *phase shifts.* Phase shifts clearly occur in nature and may even occur frequently, which would imply that the functional relationships between interacting species are often nonlinear. Nonlinear functional relationships between interacting species can also create multiple stable states and *hysteresis,* a phenomenon in which the trajectory of the stable equilibrium in abundance of the two species differs depending on the directionality of change (Scheffer et al., 2001). These sorts of simple functional relationships can trick us into concluding that two species do not interact when in fact they do, and they can leave us scratching our heads in dismay when the exact conditions that once led to some particular outcome lead to a different outcome the next time around.

All ecosystems contain many more than two species, so another dimension to understanding the complexity of ecosystem behavior is identifying precisely how these species are linked in a network of species interactions. The topology of this network is potentially far more complex than the diversity of the species themselves. To illustrate the point, consider a system with just five species. Let's assume that these five species only interact with one another as consumers and prey and that the directionality of all potential consumer–prey interactions (i.e., which species is the consumer and which species is the prey) is known. There are 10 possible consumer–prey interactions (i.e., twice as many species interactions as there are species themselves). Now let's generalize this relationship to an ecosystem containing N species. The number of potential consumer–prey interactions is $N(N - 1)/2$, or $_NP_2$. Not all of these exist or occur, but we can't be sure of the linkages that do occur in nature without a detailed observational study of the food web.

So far we've considered only direct interactions, but two species can interact indirectly when they are connected with one another by way of one or more intermediate species. For example, predators can affect plants they never eat by consuming herbivores that do eat the plants. Indirect interactions add functional and numerical complexity to ecosystems.

Let's first consider how they might increase functional complexity. In our hypothetical web of direct consumer–prey interactions, all species pairs influence one another in the same qualitative way (+/−), which is to say that consumers benefit and prey incur a cost. Qualitative relationships between species change when direct consumer–prey interactions are linked to form indirect species interactions. For example, when two direct consumer–prey interactions join to form a three-species indirect interaction, the end members of this indirect linkage both benefit through what has now become an indirect mutualism (+/+). This sort of indirect interaction, in which the effects of a consumer tumble downward through the food web, is known as a *trophic cascade* (Paine, 1980). As the food chain increases in length, the qualitative nature of the interaction between the end members alternates between an indirect mutualism and an indirect agonistic (+/−) interaction (Fretwell, 1987).

A different sort of indirect interaction can emerge between two prey species when they are eaten by a common consumer. In this case an increased abundance of one of the prey species can fuel an increase in the consumer's abundance, which in turn can cause the other prey species to decline, thus resulting in what is known as *apparent competition* (Holt, 1977). Alternatively, the increased abundance of one prey species might cause the predator to focus more intently on that prey, thus diluting the influence of predation on the other prey. There are no direct mutualisms or competitive interactions in our simple interaction web, but these qualitatively more complex interactions between species emerge through indirect interactions. The possibilities for added complexity by way of other indirect effects are vast.

Now let's consider the numerical complexity that indirect interactions can add to ecosystems. We saw above that five species, for example, can link together as consumers and prey to form 10 potential direct consumer–prey interactions. Because indirect interactions involve one or more intervening species, there are three additional levels of complexity in this particular case: three-species indirect

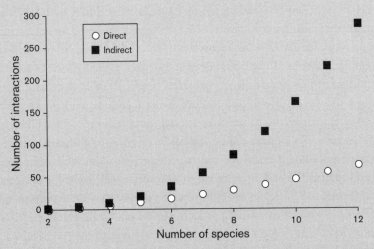

Maximum potential numbers of direct and indirect consumer–prey interactions increase with species number (graph from Estes et al., 2013).

interactions, in which the end members are separated by one intervening species; four-species indirect interactions, in which the end members are separated by two intervening species; and five-species indirect interactions, in which the end members are separated by three intervening species. As with direct interactions, we can compute the potential number of each of these indirect interaction chains. In this particular case, there are 10 potential three-species chains (computed as the number of permutations of five species taken three at a time, or $_5P_3$), five potential four-species chains, and one possible five-species chain—for a total of 16 possible indirect interactions. We can generalize this relationship for N species as

$$\sum_{r=3} {_N}P_r$$

where r = chain length. Using this simple mathematical formulation, we see that the number of potential consumer–prey interactions increases exponentially with species diversity, and of these species interactions, the number of indirect interactions quickly outpaces the number of direct interactions (see box figure). The numbers for ecosystems with more realistic species diversities are truly staggering. For instance, there are more than 2.5 billion, billion, billion, billion, billion, billion, billion, billion, billion, billion, billion, billion, billion, billion, billion, billion, billion (2.5 followed by 157 zeros) potential indirect interaction pathways in a system with just 100 species!

Chronicling the diversity of species and species interactions is just part of the challenge of understanding living nature. A more fundamental difficulty is in understanding the outcomes of species interactions and how these interactions coalesce to define how ecosystems look and operate. Not surprisingly, researchers have used a broad range of approaches in their efforts to assemble the pieces of this puzzle. The approaches can be roughly divided into two groups: theoretical and empirical. Pure theoreticians use mathematical models to characterize the behavior of species, populations, communities, and ecosystems and manipulate these models in an effort to understand the outcome of species interactions. Some theorists strive to understand ecosystem-level properties like stability, diversity, or the results of *competition* and predation. Others use theory in an effort to understand behavior. I've always been intrigued by the theoretical approach, but I knew from the outset that I would never be very good at it, and therefore I looked to the empirical side of nature for my approach to learning. Empiricism is simply inference based on observation or experiment.

Although all good empiricists are guided by theory to one degree or another, empirical approaches to understanding ecology are almost as numerous as the people who study nature. Many empiricists focus on particular species. This is especially true for charismatic organisms such as redwood trees, polar bears, condors, and pandas. Others are drawn to particular taxa, such as plants, birds, insects, or mammals. Still others direct their attention more broadly to particular biomes, such as deserts, grasslands, temperate forests, or tropical reefs, although many who study particular biomes further restrict their efforts to particular places. For example, a fair number of kelp forest ecologists I know have never ventured beyond California and northern Mexico for their field studies, even though kelp forests occur elsewhere around the world.

Just beneath the surface of these various approaches to understanding nature is the more integrating quest to understand process—which, in turn, requires both a conceptual model and a means of putting that conceptual model to a test. Conceptual models define the ways scientists have come to suspect or believe that nature works. And those suspicions and beliefs, while highly varied among individual scientists, never drift too far from consumer–prey interactions. That's because, in contrast to all other kinds of ecological interactions, consumer–prey interactions are fundamental to living nature as it has evolved on our planet over the past several billion years. Life on Earth would continue to exist in the absence of competition, albeit in some substantially altered way. The same might even be said of *mutualism, commensalism,* and *amensalism* (see box 1). But without consumer–prey interactions, there could be no life on Earth beyond the *autotrophs*—those photosynthetic and chemosynthetic organisms that obtain their energy and nutrients from the physical and chemical environments and thus form the first link in every food chain.

Consumer–prey interactions are plus–minus: consumers benefit while prey incur a cost (+/−; see box 1). There is a large divide in the field of ecology between people with differing views on whether pattern and process in nature are dictated more strongly by the "plus" or by the "minus" side of this dyad. Emphasis on the plus side implies that the distribution and abundance of species are determined by the things that nourish them—water, sunlight, and nutrients in the case of plants, and the abundance and quality of prey in the case of animals. According to this view of nature—termed *bottom-up forcing* (or bottom-up control)—the distribution and abundance of species across the planet are dictated largely by autotrophic fixation of energy and nutrients into organic compounds and by the efficiencies whereby that energy and those materials are transferred upward across increasingly higher *trophic levels* as species are consumed by the things that eat them. This is an appealing view of nature because it must be true at some level. That is, without energy and nutrients, nature as we know it would cease to exist. The logic is as simple as that.

The minus side of the consumer–prey dyad implies that the distribution and abundance of species are determined by the things that eat them (i.e., *herbivores* for plants and predators for animals), a process known as *top-down forcing* (or top-down control). The logic underlying this top-down view of nature is more tenuous because it doesn't have to be true. That is, nature could easily work just fine without consumers having any limiting influence whatsoever on the things they eat. The logically more compelling aspect of top-down control lies in the strong difference in fitness consequences between taking a bite of something that's nutritionally beneficial versus being eaten. Bottom-up effects on an organism's fitness depend on a lifetime of getting enough of the right things to eat. Top-down effects on an organism's fitness hinge on that brief moment of being captured, killed, and eaten. The logical appeal of top-down control is not that it must occur but, rather, that being eaten has an inevitably rapid and powerful influence on an organism's fitness. To the degree that populations are influenced by the fitness of their individual members, top-down forcing has a powerful influence on the workings of nature.

No credible ecologist believes that nature is regulated exclusively by either bottom-up or top-down forcing; the evidence for both is beyond question. But most ecologists lean in one direction or the other, sometimes strongly. Because of its logical primacy, bottom-up forcing is often invoked with little or no concern for top-down forcing. Open-ocean ecology developed very much out of this mindset and, in many circles, remains deeply entrenched in the same view today. On land, the subdiscipline of ecosystem ecology holds a similar perspective. The more applied disciplines of wildlife and fisheries management seem to have inherited this same view of process, except for the recognition that human exploitation influences the dynamics of populations. The proponents of top-down forcing acknowledge that bottom-up forcing exerts a powerful influence on nature—not doubting, for example, that the most striking and important differences between a tropical rainforest and a subtropical desert are due primarily to differences in the amount and seasonality of rainfall. But their minds are open to the possibility that strong top-down controls have fundamentally changed the way the world looks, often through *direct* or *indirect effects* on plants or other organisms near the base of food webs. The same open-mindedness is often lacking in the most zealous advocates of bottom-up forcing, though many of them will acknowledge top-down effects. But their worldview doesn't require top-down effects, and I suspect that in their souls, they don't believe such effects really matter very much.

Why is it that such a fundamental yet simple question as the relative importance of bottom-up and top-down forcing remains so poorly resolved? There are lots of reasons, including the inherent tendency of people to continue to believe what they already think is true. But beyond that difficulty, the only way to understand these processes with reasonable certainty (at least in my opinion) is to manipulate the purported driver of process. Seeing and understanding the consequences of bottom-up forcing requires the addition or removal of the key limiting resources at the base of the food web. Similarly, seeing and understanding the consequences of top-down forcing requires the addition or removal of consumers near the top of the food web.

Experiments of this exact nature have been done again and again, often with clear and unequivocal results. In John Martin's well-known "IronEx experiment," a pulse of elemental iron was added to the eastern tropical Pacific Ocean to test the hypothesis that iron was a limiting nutrient in ocean production; the iron produced a strong pulse of new production, thus demonstrating the importance of bottom-up control beyond reasonable doubt.

In Bob Paine's equally well-known "*Pisaster* experiment," predatory *sea stars* were removed from the rocky *intertidal zone* along the outer coast of Washington State to test the hypothesis that predation was a key process in community structure; the removal produced a striking change in the distribution and abundance of species, thus demonstrating the importance of top-down control beyond reasonable doubt. No one questions these results. Instead the uncertainties revolve around three other issues.

One of these issues is the relative importance of bottom-up versus top-down control. Seldom have these two forcing processes been looked at simultaneously or even in the same system. Work by Steve Carpenter and colleagues in Upper Midwest lakes is an important exception. These ecologists added various nutrients to lakes or removed predators from them and found that both had striking impacts on the distribution and abundance of plankton (Carpenter et al., 2001).

A second point of frequent debate is the extent to which experimental findings can be generalized across the range of the ecosystem in which the species of interest co-occur. Experimental ecological studies are labor intensive, time consuming, and therefore necessarily done in one or a few locations. Once a study has been done well and the results published, the intellectual motivation as well as the ability to obtain financial support for repeating the work rapidly wane. In the relatively infrequent instances that an experiment is repeated elsewhere, the results almost always differ to one degree or another.

A third point of uncertainty is the extent to which results obtained from a particular ecosystem apply to other systems. If the importance of bottom-up or top-down forcing processes can be generalized from one system to another, this would seem to be most valid between systems that are structurally similar to each other. However, the people who study and think about these similar but different systems often have radically different views of the controlling process. The contrast between lakes and the open ocean is a clear case in point. Lakes and oceans are similar in that both contain the same general kinds of organisms (*phytoplankton, zooplankton,* and fish) living in the same kind of physical environment (water). Despite these similarities, lake and ocean ecologists have radically different views of the ways in which these two systems work. Lakes are seen as being run by a mix of bottom-up and top-down forcing, whereas the open ocean is seen as being largely subject to bottom-up control. Are these systems really so different from one another, or is it simply that the people who study them have looked for different things? Although arguments could be made for both explanations, and lakes and oceans must differ from one another in many ways, the truth is that we don't know the answer to this question. The diversity of species and ecosystem types across the planet make the resolution of this uncertainty a truly daunting challenge.

Another obstacle to the integrated understanding of bottom-up and top-down forcing is that the experimental manipulation of would-be *ecological drivers* is more difficult in some systems than in others. In defense of the ocean ecologists, whom I have charged with having a monolithic view of bottom-up forcing, they do work in an environment in which many of the potential controlling drivers are difficult to manipulate experimentally. This feature of open-ocean ecosystems sets them apart from lake systems, which, because of their comparatively small size and well-circumscribed physical boundaries, are relatively easy to manipulate experimentally.

The longevity of individuals within populations is also a key factor in determining the degree to which purported drivers of bottom-up and top-down forcing can be revealed through purposeful experimentation. This is especially so for plants, whose generation times vary by up to six orders of magnitude (e.g., from hours to a few days for phytoplankton, to centuries and possibly even millennia for some long-lived trees). The intrinsic deterrent to doing experimental population and community ecology with species like redwoods is obvious, because 20 or more generations of scientists would be needed to see the experiment through to completion. The same thing can be done in lake systems over the course of a single summer field season.

Bottom-up forcing processes necessarily begin with plants, and while some plant species are difficult to manipulate experimentally (e.g., the long-lived trees), many others are not. Because of this, bottom-up forcing processes can be studied effectively by using the experimental method. Top-down forcing processes necessarily begin with apex predators, which as a group are more difficult to manipulate experimentally. Although consumers have been manipulated in numerous studies to explore top-down forcing processes, nearly all the well-designed and well-executed examples involved small-bodied, sedentary species of intermediate trophic status. Except for parasites and pathogens, the true apex predators are large-bodied vertebrates, most of which are also highly mobile, relatively rare, and often secretive in their behavior. Moreover, many of these species have long since been removed from their natural habitats or are so depleted that any semblance of their ecological function in the modern world is doubtful.

Other than humans, large apex predators are the most revered life-forms, which adds to the difficulty of manipulating their populations for the sole purpose of learning. Such experiments are socially unacceptable or illegal in most parts of the world. This leaves us with just three possibilities for exploring the roles of apex predators in a manner we can trust to tell us the truth. We have to watch the associated ecosystems (1) following predator reintroductions that are done for management and conservation purposes, (2) as depleted predators recover fortuitously or because of some management action, or (3) as the predator species that remain continue toward extinction. Populations of large predators can decline rapidly, but more often their declines are gradual and insidious. Population recoveries are necessarily gradual, owing to the low reproductive potential of most large vertebrate predators. For these reasons, predator ecologists often use the *space for time* approach in attempting to see and understand the roles of predators in nature, a method identical in principle to the way in which plant ecologists have studied succession in systems with long-lived species. In the case of succession, the approach is to compare different places that have existed for varying lengths of time following a major disturbance or the creation of new habitat. In the case of predators, the approach is to compare places where the predators presently occur with otherwise similar places in which the predators once occurred but are now absent. None of these approaches make for perfect experiments, but often they are powerful quasi-experiments in that they provide a dynamic view of the ways in which large predators affect their associated communities and ecosystems.

If top-down forcing processes are deeply important and widely occurring in nature, what might we expect to see when we do these quasi-experiments? Most directly and obviously, we should expect a change in primary prey abundance. If the apex predator has been reintroduced, its prey should decrease. Moreover, if top-down forcing processes occur more broadly through the food web, we should also expect to see increases in the prey's own prey as indirect manifestations of the reduction in primary prey abundance. Eventually, we should see differences in the abundance and species composition of plants as the indirect effects of apex predators spread downward to the base of the food web. These indirect effects by apex predators on plants should thus differ with food-chain length. For example, in a simple, two-trophic-level food chain with just herbivores and plants, more herbivores (the apex consumer) will result in fewer plants; whereas a three-trophic-level food chain in which predators (a new apex consumer) of the herbivores have been added, more predators will result in more plants. More generally, the indirect effects of predators on plants should be positive in odd-numbered food chains and negative in even-numbered food chains (Fretwell, 1987). And there is no reason to think that the influences of apex consumers should stop with plants. Plants, after all, are the ultimate source of nutrition and habitat for nearly all consumer species. Therefore, one might expect to see the influence of apex consumers across all of nature, wherever one looks. Every practicing ecologist has a somewhat different worldview, depending on the things they have seen, the things they have read or heard from other ecologists, and finally the deeply complex nature of their personal makeup, which dictates whether or not an idea or an argument rings true. The resulting range in beliefs and perspectives is wide indeed. Some of us see the world in fairly similar ways and some in quite different ways. Some of us approach the quest for understanding in similar ways and others in quite different ways.

I have chosen to begin with this synopsis of my own worldview of ecology because it establishes a context for the work I have done on sea otters and kelp forest ecosystems over the past 45 years. Looking back now, I really can't say to what extent this view grew out of what I learned from the sea otters and kelp forests, to what extent my learning about sea otters and kelp forests was motivated and influenced by things I learned from other people, and to what extent the approach I have taken in the quest for this understanding was influenced by my own psychological infrastructure. All these elements of my worldview of science and nature have changed over the years. The path of my research program was shaped by an ongoing interplay among these dimensions of belief, learning, and understanding on one hand and opportunity on the other.

THREE

The Aleutian Archipelago

MOST OF MY WORK ON SEA OTTERS and kelp forests was done in the remote central and western Aleutian Islands. This is due in part to the serendipitous event that led me there in 1970. Had it not been for that early opportunity, I doubt that I ever would have visited or even thought much about the Aleutians. But once there, I was drawn back again and again by two forces. One of these was the opportunity for learning. Islands are wonderful natural laboratories because they often differ from one another in ways that provide a view of ecological process. Darwin's famous journey to the Galápagos Islands helped him see and understand evolution through differences among islands in the form and behavior of closely related species. In a similar manner, the influences of human history left specific islands of the Aleutian archipelago different enough from one another for me to see and understand ecological processes that otherwise would have been invisible and incomprehensible. Strategic comparison of islands is the methodological underpinning of my effort to understand the interplay between sea otters and kelp forests. I'll return to that approach with more specifics in many of the chapters that follow.

My second reason for returning to the Aleutians whenever I could was a love for the place (figure 3.1). I have always been drawn to wild places, and no place in the modern world is much wilder or more remote than the Aleutians. Most of the islands are uninhabited and appear superficially as though they always have been. There are no people, no roads, no buildings—just tundra, rock, snow, the surrounding ocean, and the wild things that live there. The Aleutian archipelago has been a place for me to go throughout life to nourish and revitalize my soul. These two motivating factors—a love of place and the opportunities it provided for learning about nature—worked in lockstep. No

FIGURE 3.1. A landscape from Tagadak Island (central Aleutians) on a rare clear day. The M/V *Tiglax* (vessel at right center) is 121 feet in length.

matter how interesting the work, I doubt that I would have continued to return to the Aleutians were they not so viscerally appealing to me. And no matter how appealing they were to me, I surely would not have been able to return without a more concrete reason to be there.

The Aleutian archipelago, which includes more than 300 named islands, extends southwestward from the west end of the Alaska Peninsula to the east coast of Kamchatka, thereby forming what is sometimes thought of as a porous border between the North Pacific Ocean and southern Bering Sea. The islands were formed in the mid-Eocene by volcanism as the northern margin of the northward-moving Pacific plate collided with and plunged beneath the North American plate (Jicha et al., 2006). Because of their volcanic origin, most of the islands are composed of basalt, or of sedimentary rock in a few rare places where the overlying seafloor remained intact atop a growing island.

Tropical or subtropical conditions existed in the Aleutians until the Pliocene onset of the present *glacial age,* about 3–5 million years ago. As the poles cooled, the *biota* of the Aleutian archipelago gradually changed to one that was boreal or subarctic in general nature. The advance and retreat of sea ice and glaciers repeatedly scoured the Aleutians during the Pleistocene, wiping the islands clean of most species of terrestrial and shallow marine life and resetting the cycle of reinvasion by species from eastern Asia and western North America. As a result of these processes, the Aleutian archipelago contains only a few endemic species, which are quite similar to their Asian or North American progenitors. Both on land and in the shallow surrounding seas, the naturally occurring biota is highly similar across the archipelago, owing partly to the short time available for evolutionary change and partly to the uniform climate. Overall, species diversity is low, as is the diversity of habitat types within and among the islands. There are remarkably few invasive or exotic species, probably because the naturally occurring species are themselves recent invaders from Asia or North America.

All the aforementioned patterns contrast with those of lower-latitude islands around the world, which, as a general rule, are characterized by higher diversity and extreme endemism because their biotas were not wiped clean by Pleistocene glaciers and the resulting much longer periods of climatic stability. Lower-latitude islands are further distinguished by numerous invasive species, owing partly to high levels of endemism and, hence, the inevitable tendency for any invader to differ from the native biota. Finally, most large islands at lower latitudes, because of the invariant prevailing direction of tropical or subtropical trade winds and subtropical or temperate westerlies, are typified by extreme local variation in rainfall, habitat, and species composition. These various conditions complicate and confound the utility of interisland contrasts for making ecological inferences among lower-latitude islands. By comparison, the Aleutians are simpler, more uniform, and easier to use for making ecological inferences through interisland comparisons.

Although the Aleutian Islands appear to be pristine and free from human influence, they are not. The islands were first inhabited by humans when the Mongoloid peoples that initially colonized North America—from Asia, by way of the Bering Land Bridge at the end of the Pleistocene—spread southward and westward along the Alaska Peninsula. The expansion of these early humans westward across the Aleutians was apparently slowed by the expanses of open sea between the islands—because, while the eastern Aleutians were first peopled 8,000–10,000 years ago, humans did not reach the western Aleutians until some 5,000 years later. The Commander Islands, westernmost of the Aleutian archipelago, apparently were never reached by the westward spread of early peoples along the Aleutians.

Because of the unproductive land, a productive surrounding ocean, and the mild and seasonally uniform temperatures, aboriginal people of the

Aleutian archipelago—the Aleuts—were able to exist in large numbers by developing sophisticated maritime technologies and cultures. These people prospered by exploiting marine invertebrates, fishes, birds, and mammals. A history of their culture and ecology is preserved in numerous *midden* sites. These midden-based records differ from those of other maritime peoples who lived along the continental margins of the North Pacific Ocean, in that they capture the times of first contact with a previously human-free environment. This is especially true for the central and western Aleutians. Elsewhere, because of the longer periods of human occupancy and post-Pleistocene sea-level rise, midden records of initial human contact with nature have been lost beneath the sea.

Like aboriginal human cultures almost everywhere, the *Aleuts* were devastated during the period of European exploration and conquest. This happened in the Aleutians following the Bering Expedition in 1741. Although members of the Bering Expedition had only superficial and limited contact with the Aleuts, their discovery of abundant sea otters and fur seals after being shipwrecked and overwintering in the Commander Islands initiated the Pacific maritime fur trade. As fur hunters pushed eastward across the Aleutians in quest of sea otter pelts, they decimated both Aleuts and sea otters. By the early 1800s, sea otters had become so depleted that the Russians began introducing arctic foxes to Alaskan islands (the Aleutians and elsewhere) in an effort to maintain their economic value. This venture was insufficient to justify a continued presence in the New World, and in 1861 Russia sold Alaska to the United States. The United States purchased Alaska, in part, because they foresaw value in the area's natural resources, including the increasingly valuable pelts of the remaining sea otters. The mid-1860s thus marked the beginning of a period of renewed exploitation of sea otters in the eastern North Pacific Ocean, and by the early 1900s sea otter populations had been reduced to the brink of extinction throughout their range (Kenyon, 1969). By 1911, when further takes were prohibited by the International Fur Seal Treaty, many thought that sea otters had already become extinct.

But the sea otters had not quite been hunted to extinction. Although they were indeed absent from most of their historical range, a dozen or so small colonies survived here and there at remote places across the Pacific Rim (Kenyon, 1969). Several of these remnant colonies dwindled to extinction, but others survived and began to recover. Three such colonies survived in the Aleutians, one at Amchitka Island, one in the Delarof Islands 200 miles east of Amchitka, and one in the far eastern Aleutians. When Olaus Murie surveyed the Aleutian Islands in 1936 and '37, one of his purposes was to document the extent to which these remnant sea otter colonies had recovered some 25 years after the protection they had been afforded under the Fur Seal Treaty. Murie's published account of the survey (Murie, 1959) makes particular note of the location and number of sea otters he observed during his journeys across the Aleutian archipelago. Extinction was no longer a serious concern by that time. Recovery from the fur trade was well under way.

World War II was a time of increased human presence in the Aleutians because the Japanese occupied Attu and Kiska; the United States first fought to expel the Japanese from Attu and Kiska and later began major staging efforts at Amchitka and Adak for the invasion of Japan (Garfield, 2010). Smaller outposts were established at numerous locations across the central and western Aleutians in support of this activity. Some of the military and support personnel assigned to these outposts took interest in the local wildlife, and a few maintained detailed records of sea otters. Decades later, Karl Kenyon (1969) assembled these various records, along with results from his own surveys during the late 1950s and early '60s, to provide a comprehensive, updated assessment of the distribution and abundance of sea otters—thereby setting the stage for the start of my own work in 1970. By then, sea otters had recovered to pre-exploitation levels throughout the Rat Islands. They also had fully recovered in the Delarof Islands and eastward through the Andreanof Islands to about Atka. Sea otter populations had also recovered and spread through the Fox Islands and along the western Alaska Peninsula, although the exact distribution, abundance, and status of these eastern populations were less clear. There had been only the occasional otter sighting in the Islands of Four Mountains, which indicated that the species remained, in effect, absent from this region. Sea otters also remained absent from the Near Islands, except for a dozen or so animals that had only recently recolonized Attu Island in the vicinity of Chichagof Harbor (Jones, 1965). The presence of otters on some islands but not others resulted from the ravages of the North Pacific maritime fur trade, the fortuitous locations of the surviving remnant colonies, the sea otter's dependence on shallow coastal waters for food, and the resulting effective barriers to dispersal created by deep-water passes that lay between the various major island groups. By simply comparing islands with and without sea otters, I was able to see and document the species' influence on the ecosystem.

Removal of sea otters was not the only human impact in the Aleutian archipelago. The subsequent introduction of arctic foxes also influenced the islands (Bailey, 1995). These exotic predators were introduced to every island large enough to return a harvest and to many that were not. On a few of the islands these introductions failed. On a few others they were not even attempted because access from the sea was too difficult or dangerous. Thus, most islands had foxes but a few did not. And so, as with the sea otter, the nature of the fox's influences on their associated ecosystems could be seen and documented through the comparison of islands with and without foxes.

People also affected the oceans surrounding the Aleutian archipelago in ways that cannot be seen through interisland comparisons. One such potentially large class of impacts are those resulting from the exploitation of pelagic (open-sea) marine resources. The cool waters of the North Pacific Ocean and southern Bering Sea are highly productive and thus support an abundance of marine mammals, birds, fishes, and shellfishes. The main targets of exploitation in the Aleutians have been Pacific halibut, Pacific cod, sablefish, pollock, Atka mackerel, Pacific Ocean perch, king crab, and several species of great whales. These species have been removed in large numbers from waters surrounding the Aleutian archipelago, but—except for the Pacific Ocean perch fishery, which collapsed in the 1960s—the impacts of fishing on the various fish and shellfish species are uncertain, and productive fisheries continue to this day. Most of the research attention on these fisheries has been focused on the relative impacts of ocean climate change and of the fisheries themselves on the size and sustainability of fish stocks. Very little is known about the impacts of fisheries on other species or ecological processes. In a few cases, namely for the now endangered Steller sea lion, fishery effects have been invoked as a possible reason for a population collapse that occurred across southwest Alaska in the late 1970s and 1980s and the subsequent failure of the affected populations to recover. The scientists working on these species and issues have yet to figure out how to put that hypothesis to a reasonable test.

Fishing and whaling escalated and spread across the North Pacific following World War II, owing partly to the maritime technologies that had developed from the war effort and partly to the needs of Japan and Russia to revitalize their devastated postwar economies. In contrast to the impacts of fishing, which are uncertain, the direct effects of whaling are clearly seen in the *International Whaling Commission*'s (IWC) catch statistics. Only a few whales were taken from waters surrounding the Aleutian Islands until after World War II. This number was sharply on the rise by the early 1950s. Whale landings had peaked by the mid-1960s, then fell to near zero by the early 1970s as a result of both stock depletions and IWC restrictions.

Although the introduction of foxes, fishing, and whaling all seemed unrelated to sea otters and kelp forests during the early years of my work in the Aleutians, these events emerged as important elements of the sea otter's story as time progressed and my understanding of the system grew. I'll return to these events and how they are interwoven into the fabric of Aleutian ecology in chapters 10 through 13.

FOUR

Sea Otters and Kelp Forests

FOR MY PART, THE DECISION TO GO to Amchitka Island for two years to study sea otters was an easy one. I had little to lose and much to gain from the time and experience. If all else failed, I was bound to have a great adventure in an exotic and interesting place. The Atomic Energy Commission's Bioenvironmental Branch, on the other hand, must have felt considerable trepidation about sending me to Amchitka Island. I was about as green as one could be—inexperienced in research, untested in the field, and without a track record of success or failure. The AEC took a gamble on me for two reasons. One is that they trusted Vince Schultz (my former professor from Washington State University), and Vince was in my corner. Another was that they sought an individual who was willing and able to spend long periods in the field, living with and getting to know sea otters and their environment. They easily could have found someone else, but finding an older and more experienced person who was able and willing would be harder. The clock was ticking in the countdown to Cannikin (code name for their forthcoming nuclear test), and the AEC needed to get the project moving as quickly as possible. I was in the right place at the right time.

The AEC's next step was to determine how to pay me and administer the project. They might have hired me directly, but neither of us wanted to do it that way. Another option was to put me under contract, which is how the government typically does business of this nature. But the government does not often give contracts to private individuals, and I was not affiliated with an organization or institution that could serve as a contractor. The AEC had a larger environmental program that operated through a contract with Battelle Memorial Institute in Columbus, Ohio. Thus, another option was for the AEC to subcontract the work through Battelle. I didn't know anyone at Battelle, they didn't know me, and the AEC had just taken the sea otter work at Amchitka away from Battelle because they were dissatisfied with its performance. Clearly that was not the way to go.

As it happened, I had become friends with Norm Smith while we were graduate students at Washington State University. Norm and I overlapped for about a year as I was starting my master's degree and he was finishing his doctorate. Although Norm was a terrestrial wildlife biologist by training, he loved the ocean and had been deeply influenced by both a stint in the navy and a summer course he had taken on marine invertebrates at the University of Washington's Friday Harbor Labs. I liked and respected Norm, and when he left Washington for a faculty position at the University of Arizona, we spoke loosely about the possibility of me joining him some day for my PhD work. When I approached Norm about running the contract through his lab at the University of Arizona, he agreed to assume the role of both contractor and mentor. Norm helped me during the critical early days in writing the contract proposal and fast-tracking it through the AEC and the University of Arizona's contracts and grants office.

By September 1970 we had developed a scope of work and had a two-year contract with the AEC. My job would be to assess and report the impacts of the Cannikin nuclear test on Amchitka's sea otter population. This assessment was to be based on three measurements that would be taken twice, in the year before Cannikin and the year after it. One of the metrics was sea otter abundance and distribution. I was to develop a suitable survey technique and then conduct the surveys at Amchitka repeatedly over the next two years. Another metric would assess the test's effects through measures of the number and distribution of stranded sea otter carcasses on beaches adjacent to the test site. In total the beach survey area included about 30 miles of shoreline, which I would traverse monthly, noting the location, sex, age, and, when possible, the cause of death of each encountered carcass. The third metric assessed effects on sea otter behavior, namely what the animals ate and how much time they spent doing the things they do—mostly foraging, resting, swimming, grooming themselves, and interacting with other otters. The underlying idea was that any significant acute or chronic effects of the nuclear test on sea otters would be evident as differences in one or more of these measurements before and after the detonation.

While I knew I could do these things, I also realized I had an opportunity to do more. The AEC understood this and encouraged me to develop a basic research program; Norm Smith understood this and encouraged me to build the program into something suitable for a doctoral dissertation. The challenge was in figuring out exactly what to do. At the time, I knew almost nothing about the ocean and very little about ecology. I had learned a few things about sea otters by reading Karl Kenyon's excellent monograph (Kenyon, 1969), which was based largely on his own studies at Amchitka Island during the 1960s.

Armed with what I was able to learn from Kenyon's work and what I saw during my first month or two at Amchitka, the beginnings of a more basic study plan began to take form. Ideas came to me as I sat on the cliff tops watching and thinking about sea otters; as I walked the beaches in search of otter carcasses but also seeing myriad other things that washed ashore; and as I hiked across the tundra, thinking as I started out in the morning about what I might see in the day to come, or as I returned in the evening about what I had seen in the day that passed—all the time drawing inspiration from the many faces of the magnificent land and seascapes. I was alone within myself but never lonely, and three ideas began to emerge from my reading and pondering.

Kenyon's extensive work on sea otters, and the few studies that preceded it, were all based on observations of unmarked animals. Although information of this sort can be used to characterize patterns at the population level, it doesn't tell one very much about the individuals in those populations. This is not a trivial point. Understanding populations without knowing the individuals is like trying to understand how a clock works without knowing the various internal parts and what they do. I knew from Kenyon's studies that sea otters mainly ate shellfish, especially *sea urchins;* that females gave birth to a single young; that adult males defended territories; that mating occurred during brief consorts between males and females, but otherwise that males and females didn't intermingle; and that most of the beach-cast carcasses at Amchitka consisted of juvenile or aged individuals. I didn't know how extensively sea otters moved around or the size and shape of their resulting *home ranges;* I didn't know how frequently adult females reproduced; and I didn't know whether individual sea otters all ate the same things. In short, I knew almost nothing about the details that defined the natural history of this species and the dynamics of its populations. I realized that the development of a method for tagging and tracking individual sea otters was essential to understanding these things and that many opportunities for learning would surface if only I could figure out how to tag and track wild individuals. During those early months, this became my primary goal.

I foresaw two possible ways of marking otters: visual tags and radiotelemetry. I was dissuaded from using visual tags for several reasons. The commonly overcast skies and rainy and windy weather made for poor viewing, and hence I worried that tagged individuals, when resighted, might sometimes go unrecognized or be misidentified, thus biasing the data. Sea otters abounded at Amchitka Island in those days, and I worried that tagged individuals would be difficult to relocate among these many animals. Amchitka is a big place, and knowing nothing at the time about how much sea otters moved around, I worried that I might never see a tagged animal again. I realized that if I marked enough animals, I would surely see some of them now and then, but I was concerned about what the data would mean. The other possibility, radiotelemetry, was more seductive, in part because it would make relocation of the tagged individuals easier and more certain, and in part because other people were using radiotelemetry with spectacular success in studies of other species.

Puzzling over how to attach radios to otters, I thought about affixing them to the flipper-like hind limb but decided against even trying such a thing, out of concern that an instrument package that was large enough to be useful would encumber and compromise the animal's ability to swim and move about. Collars had been used successfully in telemetry studies of other large mammals, and this seemed like the best option for sea otters as well. For the approach to work, the collar would need to fit snugly enough around the sea otter's neck that it would not slip over its head, but loosely enough for the animal to groom and clean the fur beneath it. Fitting collars around the necks of some of the fresh dead carcasses I found on the beach, I thought it might work. The next step was to attach a collar around the neck of a living otter.

Capturing and holding sea otters was easy, but putting collars around their necks was quite another thing. Wildlife drugging and immobilization methods had not yet been developed for sea otters, so I had to work out a way to restrain the animals and attach the collars without getting my fingers chewed off. I was young, strong, and fearless but this was not something I looked forward to doing. It was hard on the animals and harder on me, as my scarred hands still attest. Nonetheless, I persevered and managed to become reasonably adept at wrangling and collaring sea otters. Although I was able to attach collars to subadult males and females and to adult females, I soon discovered that the collars would not remain on adult males, the problem being that their thicker necks prevented attachment of a collar that was both loose enough to ensure the animal's well-being and tight enough to stay on.

Adult males invariably pulled the collars over their heads within minutes to hours, no matter how tightly I secured them. Whatever I might be able to learn from radiotelemetry would not include adult males. Even so, the effort seemed worthwhile, and pressing on into the early summer of 1971, I managed to capture and collar 10 adult female otters and watch them in captivity for a few days before releasing them back to nature. All the instruments failed within a week or two of release, and I'm still not sure why. The collars were designed to corrode and fall off the otters within six months to a year, and I did observe several of the animals with their collars still attached during that period. I now know that otters dive much deeper than I realized at the time, so perhaps the instruments were unable to withstand the increased pressure. Whatever the reason, my attempt to tag and follow wild sea otters was a complete and utter debacle.

Although I had failed in this early endeavor, I learned two important lessons from the experience. One was to be practical. Risks are sometimes worth taking, but field studies in natural history and ecology have to be built on methods that work. The other was to ask interesting questions. In retrospect, I realized that I had become enamored with a glitzy approach and that whatever I might have learned would have been from a post hoc search for patterns in the data. I was certain I would have learned something, but I also realized that I had embarked on the tagging and telemetry study without a question. That didn't matter in the end. But had I succeeded in developing a telemetry system for sea otters, I doubt that it would have led me down such an exciting path as the one I eventually took.

A second idea that emerged from my early ponderings involved the form and function of sea otters. As an undergraduate at the University of Minnesota and a graduate student at Washington State University, I had taken several classes in physiology and was intrigued by how marine mammals were able to stay warm in the cold aquatic environment and to conserve oxygen during long, deep dives. As I delved further into the literature, I began to realize that sea otters could provide interesting new perspectives on these exciting lines of study. All prior work on the diving and thermoregulatory physiology of marine mammals had been done on *pinnipeds* and *cetaceans,* animals whose ancestors radiated from land to sea long ago. The morphological and physiological adaptations by these groups were spectacular but not surprising, given the radical nature of the transition they had to make between life on land and in the sea, and the long time they had had to do this. It struck me that sea otters were interesting because they faced all the same challenges but their ancestors were more recent expatriates from land; thus, they'd had less time to evolve adaptations in response to the challenge of marine living. Some of the consequences of the sea otter's recent link to life on land were obvious. For one, they look more like a terrestrial animal than a marine mammal. That is, they are comparatively small; they lack the blubber layer that typifies pinnipeds and cetaceans; they continue to rely on their fur as a means of insulation against heat loss to a cold environment; and their limb structure is more like that of a terrestrial carnivore than like those of the other marine mammals, whose limbs have either transitioned into flippers or been lost to a tail fluke for propulsion in the aquatic medium.

Although the possibility of studying sea otter physiology intrigued me, I lacked both training in this area and the equipment and instrumentation needed for physiological studies. So I wrote a letter to Professor Larry Irving at the University of Alaska's Institute of Arctic Biology, asking his opinion on the value of physiological work on sea otters and if he might be interested in collaborating with me. Irving was famous, a pioneer in the study of environmental and comparative physiology, and I wondered if he would even write back. But he did respond, with a long and gracious letter encouraging me to pursue this line of research and inviting me to visit Fairbanks to discuss the possibilities in further detail.

Irving was in his mid-seventies at the time and was no longer actively doing research. He thus introduced me to one of his younger colleagues, Peter Morrison, with whom I began to plan a study on the metabolic and thermoregulatory physiology of sea otters. The collaboration with Morrison was successful, but my role became one of managing and handling the animals rather than asking questions or analyzing and interpreting data. It might have been different had I arrived on the scene a decade earlier and worked directly with Larry Irving, in which case I might have become a physiologist and spent my time at Amchitka in pursuit of answers to physiological questions. As it turned out, that didn't happen and my mind remained open to the need for another line of inquiry. This led me to ecology.

I was intrigued by the kelp forest ecosystem from my earliest days at Amchitka. I loved to dive, and diving around Amchitka Island was magnificent. Diving conditions were at their very best between winter storms, when cold temperatures and diminished light combined to create visibilities in excess of 200 feet. The underwater cliffs and valleys were festooned with kelps and various other *macroalgae* and interspersed with exotic-looking fish and invertebrates, almost none of which I could identify in the beginning.

But other people at Amchitka knew these plants and animals, so it wasn't long before I did as well. Pete Slattery, a technician with Fisheries Research Institute at the University of Washington and a terrific naturalist, was especially helpful in introducing me to this new world and the species that lived there.

In the beginning I dove mostly for fun, the excuse otherwise being to have a first-hand look at the environment where sea otters hunted and procured their food. But as I observed and thought about this system, I was struck by how abundant the otters were and how rare the things they ate seemed to be. Kenyon had reported a sea otter population at Amchitka Island in the mid-1960s of about 2,500–3,000 individuals. On calm days with good visibility, I often counted more than 300 animals by scanning the ocean through my binoculars from a single point of land. As my population survey work progressed, I soon realized that Amchitka supported two to three times more sea otters than Kenyon's published estimates. Although Kenyon never believed me, I was fairly sure that the difference between his estimates and mine was simply a methodological artifact. Kenyon's surveys were done from a DC-3 aircraft, and I suspect that he missed a large number of individuals. This would be attributable in part to the aircraft's speed, lack of maneuverability, and limited visibility from the cockpit, and in part to the fact that otters at Amchitka spent a lot of time feeding, such that a high proportion of individuals were beneath the surface on foraging dives at any instant in time. My estimates were based on shore-based counts of various short stretches of coastline around Amchitka, simultaneous helicopter-based counts of these same stretches, and an extrapolation of the resulting count ratio to helicopter-based counts of sea otters around the entire Island.

By observing otters from shore, I knew what they ate. I was able to make that determination because foraging otters always return to the ocean's surface to consume their prey, at which point the prey can be identified. Kenyon's studies had established that captive sea otters needed to eat about 25 percent of their body weight daily. I thus knew that sea otters were abundant around Amchitka Island; that each otter had to eat a lot to fuel its high metabolic rate; but that sea urchins, the most common prey species in their diet, were small and relatively uncommon. I also knew that sea urchins ate kelp, so I reasoned that somehow there must be an important link in the ecosystem among kelp, sea urchins, and sea otters. The interesting questions at the time seemed to revolve around production and the efficiency of energy transfer upward through the food web from kelp to sea otters. To maintain what seemed like an extraordinarily high sea otter population density, I suspected that the kelps must be extremely productive, that much of this productivity passed through the sea urchins, and that the turnover and growth rates of the sea urchin population were high. I thought it important to document these rates and processes in order to understand why the sea otters were so numerous. Kenyon had seen and documented periods of high mortality and sharp associated declines in otter abundance. I therefore wondered whether the sea otter population at Amchitka was sustainable or was staging for another decline. I wasn't sure how to study and document these various things. But that's what most ecologists and wildlife biologists did at the time, so I believed that with hard work and a little creativity I could do the same.

There were loads of environmental scientists at Amchitka in those days. I became friends with most of them, especially the graduate students. John Palmisano (figure 4.1), a graduate student at the University of Washington whose research was focused on the ecology of rocky shores, was one of my closest friends and associates. We often discussed the things we had seen and learned, and what we planned to do on our respective projects. John had been deeply influenced by Bob Paine (figure 4.2), a young professor in the Zoology Department at the University of Washington and a member of John's dissertation committee. Bob also studied the ecology of rocky shores and was beginning to make a name for himself from work he was doing on the outer coast of Washington.

The growing impact of Bob's work stemmed from the way he practiced ecology and the way he was coming to understand ecological process. Bob was an experimentalist who believed that understanding process came most quickly and easily by manipulating nature. His early experiments involved the removal of predatory sea stars *(Pisaster ochraceus)* from patches of intertidal habitat and contrasting subsequent changes in the distribution and abundance of species with nearby control patches from which the sea stars were not removed. The findings from this early work were dramatic, unequivocal, and startling because few ecologists in those days thought that predators mattered much to the way their associated ecosystems looked and functioned. (I didn't know any of these things at the time.)

In exchange for serving on John's dissertation committee, Bob had finagled a trip to Amchitka in the summer of 1971. Bob's view of intertidal ecology was founded largely on what he had learned in Washington, and he wanted to know if the same sorts of things were happening elsewhere in the

FIGURE 4.1. John F. Palmisano, returning from work in the intertidal zone at Amchitka Island, circa 1972. (Photo by Charles A. Simenstad)

FIGURE 4.2. Robert T. Paine pointing to a keystone, source of the famous metaphor in his *keystone species* concept. (Photo by Robert Steneck)

world. A quick look at Amchitka, coupled with John's dissertation research, would tell him a lot about the Aleutians.

John knew that I was struggling to formulate clear and interesting questions about sea otters and kelp forests and encouraged me to talk with Bob. But I was intimidated by Bob's intellect and probably wouldn't have spoken further with him had John not pressed the issue and had Bob not sought me out one evening for a drink and some conversation. That fortuitous conversation was a turning point in my life. Bob listened patiently as I told him about sea otters and what I was trying to learn of them. He then recounted his own observations of the intertidal zone at Amchitka, places in the mussel beds that had been scraped clean, and his suspicion, based on the shapes of these scraped patches, that foraging sea otters had created them. Our talk turned from the intertidal zone to kelp forests and how otters might fit into the system. Bob listened to my account of what I had seen while diving and what I thought it might mean and then abruptly suggested a simple but radically different change in perspective. Rather than wondering how the kelp forests affected otters (a view Bob found to be obvious and uninteresting), why not explore how otters affected the kelp forests?

I had never thought to ask this question, but once Bob asked it for me, the wheels of my mind began to turn. All the same observations that had fueled my initial interest in how kelp forests influenced sea otters—the abundance of otters, the comparative rarity of their prey, and the abundance of kelp—served equally to fuel my new interest in understanding how otters might influence the kelp forest ecosystem. The difference was in the innate levels of excitement stimulated by these two perspectives. I began to realize that the bottom-up questions really weren't very thrilling. In one way or another, energy and material from the coastal ecosystem had to be finding its way from the kelp forest to the sea otters in a manner that was sufficient to support Amchitka's abundant sea otter population—that was a logical fait accompli. All I needed to do was mop up the details, though exactly how to do that wasn't obvious. In fact, I had asked the worst kind of question, one that was both boring and hard to answer. By contrast, the top-down effects of otters on the kelp forest system were anything but a logical fait accompli. These effects might be important, but they could just as easily be trivial. Moreover, the approach to resolving where the truth lay in this case was obvious and simple.

The sea otter population at Amchitka had recovered from the fur trade and had been lingering near environmental *carrying capacity* for at least several decades. This same pattern of extinction and recovery had also occurred at the six islands near Amchitka, which together form the Rat Islands group (see map 1.2 at the beginning of the book). The Rat Islands were close enough to one another to permit easy spread of recovering sea otters from island to island. About 125 *nautical miles* west of Kiska Island, westernmost of the Rat Islands group, lay another small group of five islands, known collectively as the Near Islands group, where a subtle difference in the vagaries of history had created a very different pattern in the distribution and abundance of sea otters. As in the Rat Islands, otters abounded in the Near Islands before the fur trade. But unlike in the Rat Islands, where extinction was only approached, the sea otters in the Near Islands had winked out completely. Moreover, because of the relatively great distances and deep-water passes that separated the Near Islands from the Rat Islands to the east and the Commander Islands (where sea otters had also recovered and abounded) to the west, the species remained absent from the Near Islands until about the mid-1960s, when Robert "Sea Otter" Jones discovered 12 animals near Chichagof Harbor on the northeast coast of Attu Island (Jones, 1965). Except for that nascent colony, sea otters remained absent from the Near Islands. In much the same way that Bob Paine had used the experimental removal of predatory sea stars to understand their ecological role in the rocky intertidal zone, I imagined using the extinction of sea otters from the Near Islands as a means of understanding the species' ecological role in kelp forests. This was exciting. My pulse quickened at the prospect of discovery.

All that remained was to go to the Near Islands and have a look. In addition to John Palmisano, two other graduate students came along: Phil Lebednik, a phycologist (a botanist who studies algae and phytoplankton), and Charlie O'Clair, an invertebrate zoologist. Our first decision was which island or islands to visit and how to get there. Nearly all the Aleutian Islands are uninhabited and inaccessible, except by ship. We didn't have ship support, which narrowed the possibilities to two islands: Shemya, which was occupied by the U.S. Air Force; and Attu, the site of a U.S. Coast Guard *LORAN* station. Both islands, together with Amchitka and Adak, were serviced weekly from Anchorage by Reeve Aleutian Airways for mail and occasional passenger delivery. We chose to visit Shemya because the island offered better support and easier coastal access. Although Shemya was closed to the public, the air force agreed to host our visit by providing lodging, dry storage space, and the use of a vehicle.

The plan was set and round-trip air tickets from Amchitka to Shemya were purchased. As the time approached, we assembled the necessary field equipment—an inflatable skiff and outboard motor, scuba tanks, a portable air compressor, our personal dive gear, and a few other odds and ends. The excitement for me was not so much in visiting an exotic new place but in having something more explicit to look for. Except for the absence of sea otters, I wondered if the shallow reefs surrounding Shemya would differ from those at Amchitka.

"Not the end of the world but you can see it from here" were the first words I saw as I exited the DC-6 aircraft at Shemya and entered a small building with my colleagues and a few other travelers, most of whom were in transit back to Anchorage. After the plane and passengers had departed, a surprised-looking airman asked why we hadn't left with them. When I told him we were the biologists from Amchitka, he went on high alert. We were taken to a small, windowless room and put under armed guard while the air force tried to sort out what was going on and what to do with us. My colleagues and I thought it was funny, but the guards didn't share our humor. Eventually we were taken to see the base commander, a serious-looking full-bird colonel who was clearly put out by our presence and not someone to be trifled with. The colonel glared at us from across his desk, telling us that while he had confirmed our authorization to visit Shemya, he also considered us a risk to military security and a threat to morale. But after this initial bluster he warmed and seemed to soften, expressing interest in what we were studying and offering to show us around the island.

Finally, at the end of the day, we walked to the shore for a brief look around. In the fading light, I was struck by two observations: the numerous *tests* of beach-cast sea urchins that were much larger than anything I had ever seen dead or alive at Amchitka, and a green hue to the beach sand. These were the first hints that sea otters mattered, and that I was on the track of something exciting.

The next day we were up early. John and Charlie prepared for a visit to the rocky shore while Phil and I geared up for a dive. We had to assemble and inflate the skiff, find some gasoline for the outboard motor, fill the scuba tanks, and locate a safe place to launch. The wind and sea were calm, so on this first dive we decided to simply swim out from shore. Although the water was clear, I couldn't see the seafloor until I slipped into the water and dropped below the surface. When I looked down at the seafloor, I was stunned by the vast numbers of urchins and the absence of kelp. I looked at Phil and saw what struck me as an incredulous, impish grin. I swam out into deeper water and then a short distance up and down the shore, trying to get a sense of whether what I was seeing was unusual or typical of the area. Every place I looked was the same—large and abundant sea urchins over a seafloor of crustose coralline algae with little or no kelp. After almost a year of diving at Amchitka, I immediately understood why Shemya was so different. In the absence of sea otter predation, sea urchins had increased in size and number, and the larger and more abundant urchins had eaten the kelp. This was my "aha moment," a profound realization that would set a path for the remainder of my life. I sat up most of the night, thinking and jotting down notes about what I had seen and what it meant.

My mind was buzzing with ideas, but the immediate problem was to document what I had seen at Shemya in an objective and rigorous manner. I had five days left to work at Shemya, and the notoriously unpredictable Aleutian Islands weather might turn for the worse at any time. My plan was to measure the density and size structure of the sea urchin population and the percent cover and species composition of fleshy macroalgae at three depths: 10, 30, and 60 feet. I would do this at several sites, time and weather permitting.

To sample the density and population structure of sea urchins, I fashioned a half-meter by half-meter *quadrat* from quarter-inch steel rebar. I then dropped the quadrat to the seafloor at an arbitrarily chosen starting location and placed all the urchins within the quadrat into a fine-mesh dive bag. After securing and labeling the sample, I swam a random number of kicks along the depth contour and repeated the process. I continued doing this until I had sampled 10 quadrats or 200 individual urchins, whichever came first. After completing the urchin samples, I swam up about 10 feet above the seafloor and estimated macroalgal abundance by assigning each species or species group to one of six coverage categories: 1 (0–5%), 2 (6–25%), 3 (26–50%), 4 (51–75%), 5 (76–95%), and 6 (96–100%). This general method, developed in studies of terrestrial plant communities, had been shown to provide unbiased estimates of areal cover. The urchin sampling was especially tedious, and I needed to make two dives of about an hour each to sample a single depth at a single site. We used quarter-inch neoprene wet suits in those early days, and two hours in the water was about all I could manage before becoming too cold to work effectively.

I had sampled just one site when a storm swept in and prevented me from diving for the next two days. Rather than attempt to sample a second site under what were still marginal diving conditions, I decided to use the trip's last day for shorter reconnaissance dives at several other places around Shemya. I at least wanted to know whether what I had seen and measured at the intensive sampling site was typical or atypical of the island's shoreline. I found that it was typical.

Although the trip to Shemya revealed an exciting ecological process, I didn't have enough data to tell that story in a scientifically rigorous way. I therefore began planning a second trip to the Near Islands for the summer of 1972. My overall goal was to acquire similar data from four sites at Shemya Island. I also intended to visit Attu Island, about 40 miles west of Shemya, to survey the sea otter population and determine whether the patterns I had seen at Shemya also occurred around Attu, as they should have if my model was correct. In the interim I would sample the seafloor at Amchitka. I would like to have visited a second island in the Rat Islands group as well, but none of these islands was accessible except by ship, and I didn't have the necessary ship support to make such a journey. I obtained the data from Amchitka during the remainder of the summer of 1971. I returned to Shemya in July 1972 with a dedicated diving assistant, and for 10 days we sampled the three additional sites. After that it was on to Attu.

Above the water, Attu was a different-looking place than either Amchitka or Shemya. The latter were flat or gently rolling, whereas Attu was extremely mountainous. Except for a landing strip and the coast guard's tiny LORAN station at the head of Massacre Bay, Attu was mostly a roadless wilderness. The coast guard had a single old flatbed pickup they used for hauling supplies between the landing strip and the nearby LORAN station. Coast guardsmen graciously drove us along a poorly maintained road to Murder Point, about 3 miles from the LORAN station and near a rocky promontory from which we could dive into the ocean. World War II Quonset huts provided a modicum of protection from the rain and weather. There would be no warm rooms to sleep in, no hot showers, no heated space to dry our wet suits, and no transportation except by foot. None of this mattered much. Although stripping naked in the cold and crawling into a clammy, damp wet suit weren't things we looked forward to each morning, the weather was good, the diving was excellent, and I was secure in having obtained the necessary data from Shemya and in knowing that, underwater, Attu looked exactly like Shemya. Camping on Attu, exploring the various artifacts left from World War II, and hiking the surrounding mountains were great fun (figure 4.3).

The day before our departure from Attu dawned clear and still. I decided to visit Chichagof Harbor to survey the otter population and to have a look, as best I could without scuba gear, at the associated seafloor. The Japanese had occupied Attu during World War II, and their encampment was in the hills behind Chichagof Harbor. After the battle of Attu in May 1943, U.S. forces built a gravel road from Massacre Bay through the mountains to Chichagof Harbor. After a little begging and cajoling, one of the coast guardsmen on leave for the day agreed to drive us as far as possible along that road. I still remember that drive through the mountains behind Massacre Bay and into the glacier-cut valley leading to Chichagof Harbor as one of the most breathtaking journeys of my life. The surrounding peaks were covered with snow; the rivers were wide and clear; the valley floor and adjacent hillsides had greened up from the preceding winter and were awash with color from the many arctic wildflowers; and all of this loveliness was accentuated by the rarely seen morning sunshine and a cobalt blue sky. As close to God as I would ever be, this is where I want my ashes spread when my time on Earth is done.

The rest of the day and indeed the remainder of my time in the Aleutians were enjoyable but anticlimactic. I hiked along the shore of Chichagof Harbor and the adjacent headlands and counted 25 sea otters, clear evidence that population recovery in the Near Islands was proceeding. I could see well enough into the shallow water to determine that the area was still extensively overgrazed by sea urchins and supported little or no kelp. I thought of the future, and about how interesting it would be to document changes in this system as sea otter numbers increased in the years and decades to come. But

FIGURE 4.3. The author, grilling steaks al fresco on Attu Island in 1972. (Photo by Norman S. Smith)

first things first. I completed the remainder of my fieldwork at Amchitka and returned to Tucson and the University of Arizona in late September to write my dissertation and complete my doctorate.

A scientific project is not finished until the data are published, and John Palmisano and I began discussing how to do this. I had data from the *subtidal zone* communities at Amchitka and Shemya, and John had similar information from the rocky intertidal zone. We thought that our data and story might be worthy of a report to the journal *Science,* so in the summer of 1973 John flew from Seattle to Tucson to write the manuscript with me. Neither of us had much experience with scientific writing, but between us we knew the issues and were able to complete a first draft in just several days. There were seemingly endless revisions over the next six months, after which we obtained helpful advice and input from Paul Dayton and Bob Paine. I submitted the manuscript in the spring of 1974, which went through peer review without fanfare and was published in *Science* on September 20 of that year (Estes and Palmisano, 1974).

I thought it was a good paper, but I had no idea how it would be received by the scientific community until various faculty members at the University of Arizona, many of whom I barely knew and had never spoken with, stopped by my office to congratulate me and ask for a reprint. A flood of reprint requests in the weeks that followed quickly exhausted my supply and reinforced the feeling of triumph. I felt lucky, but I also felt that I had done a good job. I now knew that I had what it took to succeed in science. I was inspired to press on, although at the time I had little hope of returning to the Aleutians or of continuing to work with sea otters and kelp forests. It was time to look elsewhere for a future.

FIVE

A Toe in the Arctic Ocean

BY MID-1974, WITH THE END of graduate school in clear sight, time had come for me to find a real job. I considered several offers but eventually accepted a position with the U.S. Fish and Wildlife Service (USFWS) to work on arctic marine wildlife in Alaska. The Marine Mammal Protection Act was signed into law in 1972, providing mandates and funding for expanded research on marine mammals and their associated ecosystems. The most common groups of marine mammals—whales, dolphins, seals, and sea lions—were assigned to the National Marine Fisheries Service, whereas the USFWS assumed responsibility for the catch-all of remaining species—polar bears, sirenians, walruses, and sea otters. This division was a consequence of the Nixon administration's earlier decision to move the two main divisions of the USFWS, the Bureau of Commercial Fisheries and the Bureau of Sports Fisheries and Wildlife, into different cabinet-level branches of government. Most species under USFWS jurisdiction were in Alaska, and for that reason I was posted to Anchorage. In late December 1974, I packed my belongings into a Volkswagen bus and once again headed north to Alaska. Two weeks later, after an eventful trip during which I very nearly froze to death, I arrived in Anchorage, excited to be back and ready to begin a new job and a new life.

Two marine mammal projects were run through our office in Anchorage, one focusing on polar bears and the other on sea otters and walruses. The polar bear project was led by Jack Lentfer, and the sea otter–walrus project was led by Ancel Johnson, my supervisor. Ancel was supervised by Clyde Jones (figure 5.1), director of the National Fish and Wildlife Laboratory. The National Fish and Wildlife Laboratory, housed in the National Museum of Natural History in Washington, D.C., was something of a modern-day

FIGURE 5.1. Clyde Jones, doing what he loved most (preparing specimens for museum collections), several years before his death in 2015. (Photo by Michael A. Bogan)

reincarnation of the Biological Survey made famous by such classic old naturalists as Olaus Murie. Although we planned to work together, Ancel's focus would be on sea otters and mine on walruses.

In the summer of 1974, before even finishing my doctorate, I spent about a month and a half traveling across arctic Alaska, meeting people and assessing the needs and opportunities for walrus research. My travels included a month on St. Lawrence Island, living with and getting to know indigenous Alaskans in the villages of Gambell and Savoonga. These people hunted walruses during the spring as the animals migrated north on the receding pack ice from their wintering grounds in the Bering Sea, through the Bering Strait, and into the Arctic Ocean. I saw the villages as potential staging areas for work on walruses.

The short-term objective of the walrus program had been established before I was hired, through a formal agreement with the then Soviet Union as part of the U.S.–U.S.S.R. cooperative program on studies of the environment. Pacific walruses range widely across the Arctic, from western North America to eastern Asia—thereby requiring a coordinated effort between the two nations for an overall assessment of the walrus population. The survey was scheduled for late summer 1975. My job was to design and conduct the survey, and then to provide an analysis of the distribution and abundance of walruses east of the International Date Line (IDL). The Soviets would do the same in the west, and we would then combine this information for the first-ever overall population assessment of Pacific walrusus. In early September 1974, I flew to Anchorage, where I met up with Ancel and Tom Balleau, the pilot of our Grumman Mallard survey aircraft. Our plan was to fly to Point Barrow, recon the ice edge of the Arctic Ocean to determine roughly where the walruses were, and use this information to delineate the area we would survey more rigorously and intensively the following year. We lived in and operated out of the Naval Arctic Research Lab at Point Barrow. I visited the lab's small museum during my first afternoon and was unnerved to see the tail section of a small plane that had crashed nearby on August 15, 1935, killing Wiley Post and Will Rogers.

A few days later we flew east to the Mackenzie River delta in search of walruses. The day began well enough, with clear air, no wind, and a high overcast that helped contrast the dark bodies of *hauled-out* walruses against the underlying ice. But as we were returning to Barrow, the overcast descended to envelope us with freezing fog. Tom tried to fly above the overcast, but it was too high. He then tried to fly below the overcast, but it went clear to the deck and we were unable to escape in that direction. We were taking on ice, suddenly in serious trouble.

Tom radioed the DEW Line stations (Distant Early Warning Line, a system of radar stations designed to detect incoming Soviet bombers during the Cold War) at Lonely and Oliktok as possible landing sites, but both reported "zero-zero" visibility. He flew low over the coastal lagoons in hopes of picking up warmer air off the water to melt the rapidly accumulating ice. I was growing increasingly uneasy as I watched fingers of ice thicken and grow along the pontoon and the underside of the wing. I noticed that the plane was beginning to drift and yawl and that Tom was steadily increasing engine power to maintain flight. I could see the water below, now covered with white caps, and hoped that Tom was not contemplating a water landing, which I knew we would be lucky to survive. At about that time the aircraft stalled and I closed my eyes in preparation for what would surely be a quick end. But Tom adeptly shoved the throttles forward to full power and we managed to remain airborne.

Moments later we passed out of the fog into clear air. But we were several hundred miles from Point Barrow and still carrying a lot of ice. More engine power was required to remain airborne, which raised the question of whether or not we had enough fuel to make it back to Barrow. Tom told us it would be close. As we approached Barrow, I watched the fuel-gauge indicator dance around zero while the low-fuel warning light flashed red. As we lined up for final approach, Tom told us to hang on because we would be landing at high speed. When we finally touched down and taxied to the tie-down, I felt what seemed like a lifetime of stress and anxiety flow out of me. I'll never forget Tom's words when we stepped down onto the tarmac and began walking toward the terminal building: "Gentlemen, that's as close as you'll ever want to come."

Later that evening I sat down with Ancel over a glass of whisky to celebrate our survival and discuss the future. Neither of us wanted to get in that plane again. But we did, and a week or so later we flew back to Anchorage, feeling lucky to be alive but otherwise little the worse for wear. I returned to Tucson to finish my dissertation and make ready for my move to Anchorage. Several weeks later I received a letter from Ancel, telling me that the Grumman Mallard that had nearly been the end of us, and that we had intended to use again the following year for the walrus survey, had gone down over the northern Gulf of Alaska during a seabird survey. There were no survivors.

Aerial surveys had become embedded in the culture of wildlife biology. These endeavors are intrinsically dangerous. Planes crashed and people died—that was just part of it all. When a plane went down and a biologist was killed, a new plane was obtained and another biologist hired or assigned to the job. The Department of the Interior's Office of Aircraft Services thus began looking for a new, and hopefully safer, survey platform. They selected a surplused Lockheed P2V, precursor to the famous P3 submarine chaser and, at the time, holder of the world record for time in the air between takeoff and landing. There was an observation bubble on the nose of the P2V, ideal for wildlife surveys. By late fall of 1974, the aircraft was in Anchorage and being made ready for use.

Walruses are highly gregarious animals. One of the challenges in counting them is the accurate enumeration of group size, which often ranges into the thousands of individuals. The P2V had a belly window for surveillance, and I thought we might solve the group-size counting problem by photographing the larger groups. After a little looking, I found an aerial photographic expert with the Bureau of Land Management who was keen to help. We mounted a large-format, high-resolution, motion-compensated camera in the P2V's belly window, hoping it would provide the images and data I required. In March 1975 we set out to test the system over the pack ice in the southeastern Bering Sea, where walruses were known to occur in abundance during late winter.

After waiting for a period of suitable weather, we departed Anchorage in the P2V early one Friday morning. Our plan was to search the frozen eastern Bering Sea until we found some walruses, at which point we would descend to an altitude of 500 feet and fly several mock survey transects over the ice, simultaneously photographing the walrus groups we encountered.

The walruses were easy to find, and we quickly began flying transects and photographing the groups. We had been at this for about an hour when I heard a muffled explosion and felt the aircraft shudder. The port-side reciprocating engine had overheated and blown up. I wasn't terribly concerned for our safety, because the aircraft was equipped with another, starboard-side reciprocating engine and outboard jet engines. My main concern at that moment was for our work and the likelihood that we would need to terminate the survey and return to Anchorage with only a limited amount of data. But in fact we were in trouble. Our pilot feathered the port-side engine, but the increased power required of the starboard engine was causing it to overheat. The jet engines thus had to be started up and used to keep the P2V airborne. This wasn't so worrying until I learned of the much higher rate of fuel consumption by the jet engines and that our onboard fuel supply, at this high rate of consumption, might not be adequate to get us to the nearest landing strip, which was Dillingham, about 200 miles to the east. Our pilot, apparently looking forward to an early return to Anchorage for Friday night and the weekend, hadn't loaded enough fuel aboard the aircraft for such emergencies. An animated discussion between the pilot and copilot ensued over the likelihood of surviving a landing on the sea ice. From what I was able to gather, this was just about zero. I knew we were in real trouble when, shortly thereafter, the pilot called in "Mayday" (a radio distress call used by aviators and mariners to indicate a life-threatening emergency).

Our only hope was to make it to Dillingham, which would take about 45 minutes if we didn't run out of fuel first. I listened anxiously for any further sign of mechanical trouble as we droned on to the east at low altitude. Eventually we crossed from sea to land and I could see the Dillingham runway in the distance. Not until we finally touched down was I sure that we would survive. We taxied to the small terminal building and climbed down to the tarmac, at which point a normally soft-spoken and reserved Ancel lit into the pilot for not having more fuel on board. Ancel was hot, and I thought a fight might come of it. But the pilot knew he was in the wrong and skulked off somewhere while the rest of us headed for the bar to recount the day's adventure and celebrate our survival. A few weeks later there was an investigation, during which one of the investigating officers confided in me that the pilot had in fact made two serious errors, one in not putting more fuel on board the aircraft and the other in not immediately taking the aircraft to high altitude, where the jet engines would have operated more efficiently.

Ancel was sensitive to the emotional drain and must have been asking himself whether what we might learn from the next year's planned walrus survey was worth the risk. After these two close calls, he asked me if I wanted to back out. I'll admit to being tempted, but I also felt a commitment to finishing the job and did not want my first assignment to be a failure. I would complete the survey in 1975, or die trying. The Office of Aircraft Services assured me that the P2V was properly repaired and could be used to complete the survey in a manner that was as safe as possible.

At about this same time, I was contacted by Jim Gilbert, a postdoctoral fellow at the University of Washington who was interested in marine mammal surveys and population assessment. Jim offered to join me in conducting the walrus survey and in analyzing and reporting the results. I needed a second person, and Jim had the necessary skills and interests. So I gladly accepted his offer, and together we began planning the survey. In early September we flew to Point Barrow to begin.

We had developed a sampling design for the survey that we thought would provide the closest thing possible to an unbiased estimate of the distribution and abundance of Pacific walruses east of the IDL. We knew that most walruses in this region occurred along the southern edge of the pack ice, and we also knew from satellite images the approximate day-to-day location of the edge of the pack ice. Our strategy was to fly north–south transects across the ice edge that extended far enough into the open water to the south, and far enough into the consolidated pack ice to the north, to include all the walruses. We didn't know how far we might have to go in each direction to do this, but we imagined that the distances might vary considerably with longitude and from day to day. Therefore, we planned to make those determinations on the spot, using a flight path 50 nautical miles long since the last observed walrus as the transect endpoint and delineating the overall distribution of walruses from these data.

To ensure an unbiased sample, the longitudes of the survey tracks were randomly chosen beforehand at the beginning of each day from a region bounded by the IDL to the west and by the border between Alaska and the Northwest Territories to the east. The sampling order was always east-to-west because that left us closer to the refueling center at Wainwright at the end of each day's surveying. Upon completing a transect, we turned and flew diagonally to the beginning of the next transect. We counted walruses, counted or estimated group size, and measured the perpendicular distance from the aircraft's flight path to the sighted animal or animals. This procedure was repeated on five days when weather conditions were suitable for flying and surveying.

We began flying on September 1 and finished on September 12. The survey came off without a hitch. There were no anxious moments, and the weather was generally good. But as the survey progressed, a number of disconcerting patterns appeared in the data. On some days we saw lots of walruses and on others we saw very few. On the days that we saw few walruses on the ice, we also saw few in the water. On one of the days that we saw few walruses, there were numerous oval-shaped holes in the newly formed ice, created by the heads and protruding tusks of walruses as they pushed through the ice to breathe after a dive—a tell-tale sign of animals in the area. On the other days on which few walruses were seen, I saw none of these holes in the newly formed ice. On days when large numbers of walruses were seen, the numbers counted were highly variable among transects, with no apparent relationship to geography. We would often count hundreds of walruses on one transect and then none on the very next. We saw large numbers of animals in some particular place on one day and then none in that exact same place a few days later. Although our sample size of five survey days was relatively small, there was no apparent relationship between the number of walruses that were seen hauled out on the ice and weather conditions. Before even analyzing the data, I knew that these various features of the behavior and distribution of walruses in the pack ice would be problematic to the development of a population estimate that anyone could have much confidence in.

When we returned to Anchorage in mid-September, I set out to analyze the data and report our findings. I looked at the data in great detail, in an effort to understand what could and could not be concluded. These analyses confirmed my suspicion that the walrus population could not be reliably estimated from these data. The survey was designed so that the data obtained on any given day could be used to estimate the number of walruses in the overall survey area, the initial hope being that these data could then be combined among days to increase sample size and thus obtain a more precise estimate of total abundance. But the difference in numbers of animals counted on different days was too large, ranging from a low of 125 on September 5 to a high of 3,213 on September 8, to allow the data to be combined. The total population estimates that I obtained by expanding the transect samples to the entire survey area ranged from less than 2,500 to more than 100,000 animals, depending to some small degree on the specific assumptions of the estimator but mostly on the survey day.

I was fairly certain that most of this variation was caused by differences among survey days in the proportion of walruses that were hauled out on the ice, but without information on the animals' whereabouts and behavior over the period of the survey, I couldn't be sure of that because the variation among numbers of walruses counted on any given transect for any given day was so great. For the survey data obtained on a given day, the mean number of walruses counted per transect was more than an order of magnitude less than the variance in the number of walruses counted among the transects. These data translated into coefficients of variation (standard deviation/mean) ranging from 1.66 to 2.44 for the different survey days. As a general rule of thumb, coefficients of variation in the range of 0.2 or less are required in order for a sample to provide a reasonably reliable representation of the population.

I did two additional analyses in an effort to improve the quality of the population estimates. The first of these was a post-stratification analysis. Stratification is a commonly used sampling method to reduce variance. The idea is fairly simple. If the *sample space* (in this case the area occupied by walruses) can be subdivided into regions (strata) in which variation within the strata is low and variation among strata is high, a more precise population estimate can be obtained by removing the variation among strata from the variance in overall estimated population size. Post-stratification estimation procedures are frowned on because the strata might themselves be artifacts of sampling error rather than real differences across the overall sample space, in which case the resulting estimates will be biased. I wasn't so concerned with this, because I knew that the data were a mess and I simply wanted to see how much the quality of the estimates might be improved if the pattern of spatial variation in our survey data were real, and if stratification across that pattern of variation could be used effectively in the design of a future survey.

Although I explored numerous potential stratification schemes and used a variety of procedures for the allocation of sampling effort among the strata, the resulting population estimates remained disturbingly imprecise. For example, I determined that an estimate that provided 95 percent confidence of being within 20 percent of the true size would necessitate sampling the high-density strata in their entirety and about half of the area in the lower-density strata. A sampling effort of such a magnitude would require about a tenfold increase in sampling effort.

My charge at the beginning of this project had been twofold: first, to obtain the best possible estimate of the number of walruses east of the IDL; and second, to assess the reliability of the method and the quality of the resulting data. I had done these things to the best of my ability. In early January 1976, I flew to Moscow to exchange the data with my Soviet counterparts and discuss the results of our overall findings. The Soviets faced a different physical situation and thus approached their survey of the region to the west of the IDL in a fundamentally different way. During most years, the walruses in this latter region are not dispersed across the pack ice but instead tend to aggregate in large numbers around shore-based haul-out areas. It was thus possible to account for some significant proportion of this western portion of the population simply by counting animals on the haul-outs. That approach provided no way of knowing how many animals might be at sea, thereby yielding a population estimate that was inevitably biased low to some unknown degree. The advantage of the approach was that tens of thousands of individual walruses used these western shore-based haul-outs, and therefore we could be certain that the population was at least that size.

I don't believe that the Soviets ever grasped the nature of the problem we faced in the east or the underpinnings of my assessment of the data. They seemed confused and even angered by my claim that we didn't know how many walruses occurred east of the IDL during the preceding summer's survey. And there was little sympathy or support for my suggestion that we carefully assess the utility of conducting the next planned survey in 1980.

About a month after returning from Moscow, I joined Francis "Bud" Fay and Howard Ferren on a 10-week cruise into the pack ice of the southeastern Bering Sea aboard the M/V *Zagoryani*, a Polish-built Soviet sealing and trawling vessel. Bud was a professor and Howard was a graduate student of Bob Elsner's at the University of Alaska. Walruses live in remote areas of the pack ice, and the cruise provided a rare opportunity to observe the animals and their ecosystem. Howard, Bud, and I flew from Anchorage to Dutch Harbor, where we were to board the *Zagoryani*. Because of restrictions on permits and a generally high state of Cold War paranoia among both the Soviets and the Americans, we had to do this on the high sea. After spending several days in Dutch Harbor, with no information on the *Zagoryani*'s whereabouts, we received word that we were to rendezvous that night at a predetermined location about 50 nautical miles to the north. We chartered a crabber and departed Dutch Harbor about 8:00 P.M. Although the weather wasn't great, it wasn't as bad as it often is that time of year, and we were able to make the transfer without incident. After introductions to the captain and the Soviet scientific crew, we began steaming north toward the pack ice in search of walruses.

I awoke the following morning to occasional scraping sounds on the ship's hull from collisions with small icebergs. As the day progressed, the seas subsided and the ice thickened. For the next 10 weeks we cruised through the pack ice, looking for walruses and seals. The cruise was mainly a commercial venture, the goal being to harvest walruses, spotted seals, and ribbon seals. When animals were sighted in the distance, several hunters with Soviet military rifles would venture forth onto the ice and attempt to get close enough for a shot. These firearms were far too light for an animal as large as a walrus, and I estimated that roughly half of the individuals that were shot escaped into the water.

Once an animal had been dispatched, it was hoisted aboard the ship for processing, and a flurry of cutting, sawing, hacking, and grinding ensued. These on-deck activities were nothing short of mayhem as the Soviet workers yelled out to one another in Russian, knives flashed, body parts were dragged across a bloody and slippery deck, and the sound of the below-deck grinders provided a macabre backdrop to the grisly scene. I learned that the skins would be used to make fishing trawls, that the blubber would be rendered into soap, and that the rest of the carcass would be used as mink food back in Siberia. Week after week, our days were spent in this endeavor. This was my first direct encounter with walruses, and I was along mainly to observe and learn. It's a good thing, because I was permitted to leave the ship on one day only, and then for just several hours.

The ice was constantly changing. Some days were clear and cold, others were overcast and warmer. On the cold days I watched the new ice form as oily-looking fingers across the leads of open water. Intense late-winter storms periodically swept eastward across the North Pacific Ocean and Bering Sea, often creating large swells that rolled through the ice from the south. The flexible sea ice bent above the smaller and longer-period swells, creating the bizarre perception of a rolling landscape. The larger swells broke up the pack ice, creating leads and areas of open water. On several occasions, during prolonged periods of cold, still weather, we became stuck in the frozen sea until the next set of storm-generated waves came along to break up the ice and set us free.

Various seabirds concentrated around the leads, sometimes in large numbers. Long-tailed ducks, king eiders, and common eiders were numerous, as were common murres and black-legged kittiwakes. On occasion I was treated to a glimpse of a species I had never before seen, such as ivory and Ross's gulls. On some days the weather was calm, and on others the wind howled with terrific force. As the season progressed from late winter to early spring, the days lengthened, and the sunlight intensified, ocean color shifted from clear to a murky green while long sheets of diatoms formed on the undersides of the sea ice. I wondered how these various species and patterns of change fit into the dynamics of the food web and whether processes occurred in this system that were as intriguing and powerful as those I had seen several years earlier among the sea otters and kelp forests. I was sure there must be some interesting biology going on, but I had no idea what that might be or even how to look for it.

We departed the ice in early May. For reasons I never knew, the captain was reluctant to return to Dutch Harbor and therefore decided to disembark us in Anchorage. As we traveled south from the ice into the more open water of the southern Bering Sea, the seas began to build and the ship started to pitch and roll. After two and a half months in the calm of the ice pack, we weren't used to this and everyone felt at least a little seasick.

By dinnertime I was feeling marginally well enough to eat and thought I had better have some food. It had become very rough and there wasn't another soul in the mess, except for the young woman whose job it was to serve the meals. Apparently she had been thrown against the wall of the pitching ship, because she had a nasty-looking bruise on her forehead and one of the lenses of her glasses was shattered. The poor soul sat on a stool, vomiting into a bucket between her knees. She managed to bring me dinner, but the sound of her constant retching and the smell of the cooked cabbage were more than I could take. I bolted for the door and the nearest receptacle before losing it.

Early the following morning, as we cleared Unimak Pass and turned east toward Anchorage, we encountered the full force of the North Pacific in early spring. The wind was southeasterly at about 50 knots, and the seas had built to about 35 feet. Conditions continued to worsen, and by the time we reached Kodiak the seas had built to 50 feet. We were quartering into the weather, which created a violent pitch and roll. It was so rough that I could barely stay in my bunk, much less sleep. On several occasions I went up to the wheelhouse to watch the spectacle of the monstrous, windblown waves and the occasional surge of green water over the bridge. By the time we reached the more protected water of lower Cook Inlet, I was exhausted from lack of sleep and just hanging on. After 10 weeks in the ice and three days of pounding our way home, I was relieved to set foot on dry land and revitalized by the smell of early spring in Anchorage.

Although the way forward for work on walruses was uncertain, one direction not to take was clear to me. I didn't think we should spend another dime or moment of time on walrus surveys, unless and until a means could be developed to reduce the variance of the resulting population estimate. Otherwise I felt strongly that the effort was unjustified, given the cost, the risk, and the limited value of the resulting information. I published the details of the 1975 survey in 1978, including this assessment (Estes and Gilbert, 1978).

The results seemed definitive, and I thought this would be the end of the walrus aerial surveys. But it wasn't the end. Although no one has ever formally challenged our results or conclusions, the USFWS has continued these surveys in one form or another at roughly five-year intervals ever since, spending millions of dollars along the way and at considerable risk to human life. In my view they are no closer now than they were in 1975 to knowing how many walruses there are in U.S. waters or what the trends might be for the population.

The survey results, together with my experiences at St. Lawrence Island and on the *Zagoryani,* gave me reason for pause in thinking about further research on walruses. The important overarching questions for both basic and applied research—such as the status and trends of the walrus population, how walruses might be affected by changes in the ocean ecosystem, and how walruses influenced their local ecosystem—struck me as being difficult to answer. Had this been my first foray into wildlife research, I might have simply accepted the situation as "the way it is." I might have been intrigued enough with the magnificence of the animals and the wonders of the High Arctic to press on in one way or another, knowing down deep that I would probably never learn much of any real importance about the animals or the ecosystem—but also thinking that that was okay. However, my earlier work on sea otters in the Aleutians had changed me forever by making me understand that it didn't have to be that way, that there were interesting and important things to be learned about animals and their ecosystems. I wondered if someone else might be better suited for working on walruses—and was quite sure that I was better suited for, and would be happier doing, something else. Ancel was sympathetic, but I know he was also disappointed because he felt that it was our job to develop some kind of a research program for walruses.

I had wanted to return to Attu to follow the expected recovery of sea otters and the ecosystem, and Ancel agreed, so long as I agreed to continue in some way with walruses. Clyde Jones was much more emphatic. He had hired me because of the work I had done on sea otters and expected more of the same. If that wasn't going to happen with walruses, then he wanted me to do something else.

I continued working on walruses for a number of years, but I never did another aerial survey and I never returned to the High Arctic. All of my subsequent work was done at Round Island, in eastern Bristol Bay, where thousands of adult male walruses hauled out during summer and fall after the seasonal pack ice had receded into the Arctic Ocean. From its beginning in 1976, the Round Island project was a collaborative effort with the Alaska Department of Fish and Game. The state was concerned about protecting these animals from harassment and poaching. My interest was in better understanding the comings and goings of the walruses as they moved between land-based haul-outs and foraging areas in the surrounding ocean. I hoped, in particular, that we might use Round Island to develop a tagging and tracking system for walruses. Except for earlier work by Edward Miller, a student of Bud Fay's, almost nothing was known about walruses at Round Island.

In 1977 the fish and game department hired Jim Taggart as a seasonal employee to watch over the walruses at Round Island. I hired Jim's wife, Cindy Zabel, to accompany him and help with the research. Cindy and Jim were superb—bright, creative, and as tough as they come. For several years they traveled to Round Island in May as the walruses were gathering, and returned home in September as the animals left to migrate north, to meet up with females and the southward-forming seasonal pack ice.

Several important discoveries were made during these years, due in large measure to Jim and Cindy's abilities and efforts. By conducting regular daily surveys, they found that walruses dispersed en masse to the ocean and returned to haul out on Round Island on about a two-week cycle. Upwards of 15,000 walruses packed the beaches of Round Island during the high points of these cycles, whereas only a few hundred animals remained during the low points. Jim and Cindy found that while weather conditions influenced the animals' distribution on shore, their comings and goings between land and sea were largely independent of weather. These findings were reminiscent of the high day-to-day variation in our 1975 aerial survey results.

We had no information on the behavior of individual walruses, and our efforts to mark animals using paint and other visual tags were unsuccessful, in large measure because of the great number of animals at Round Island and their tendency to pack so closely together on the beaches that the marks often could not be seen, even when the marked individual was in plain view. Radio-transmitter tags would be needed to more reliably relocate the tagged individuals. Tusks were an obvious attachment site for radio transmitters. Initially we tried doing this by drugging the animals and affixing the transmitters to their tusks while they were immobile. Although we successfully marked nine individuals using this approach, the animals proved sensitive to immobilizing drugs and we abandoned drugging after killing a second walrus.

The tagging technique was nonetheless promising, and during the following winter Jim began working with a machinist at the University of California, Santa Cruz, to develop a tag-attachment tool. The eventual product of this effort was remarkable in both its effectiveness and its mechanical complexity. The tagging tool, powered pneumatically by a standard scuba tank, rapidly compressed and then crimped a stainless steel hose clamp and the affixed radio transmitter around the sample tusks we had brought to Santa Cruz for testing. But attaching these devices to living animals in the field was another matter. The tool was big and heavy. Walruses are large, and potentially dangerous when alarmed at close quarters. The tags would have to be delivered by someone strong enough to haul the tool over Round Island's rugged terrain, stealthy enough to approach a sleeping walrus without alarming it, and agile enough to make a quick retreat after the tag was delivered. Jim had all these qualities, and with patience he was able to tag a small number of walruses during each of the next two summers.

The tagging technique was spectacularly successful. Tags remained on the walruses' tusks for months and delivered detailed information on their comings and goings from Round Island. This work was done in the mid- to late 1970s, before the advent of laptop computers and field data-entry techniques. We had obtained a large volume of information from three summers of fieldwork, but the analyses proceeded slowly as Jim finished his dissertation, went through a personal crisis when his marriage to Cindy fell apart, and then took a research position with the National Park Service in Glacier Bay. Unfortunately, all of Jim's data were lost when his house burned to the ground, before they had been carefully analyzed and published.

For my part, the draw of sea otters and the Aleutians remained close at hand. In August 1975, I returned to Attu to reassess the distribution and abundance of sea otters. My hope beforehand was that the population at Chichagof Harbor had continued to grow and spread, and that I might use further population growth in years to come to detail the patterns of population recovery and to chronicle changes in the coastal ecosystem that I expected to follow.

I flew to Shemya, where I met up with the M/V *Aleutian Tern,* a 65-foot vessel owned by the Alaska Maritime National Wildlife Refuge and used in support of their various activities in the Aleutian Islands. The *Aleutian Tern* took me to Chichagof Harbor, where I was left with some camping gear and an inflatable skiff. They would return for me in a week, after which we would circumnavigate Attu to survey sea otters and other marine wildlife. The weather was fine, and I spent the next several days surveying sea otters from Sarana Bay in the east to Holtz Bay in the west. The sea otter population had increased to about 100 individuals. I was able to travel far enough in the inflatable to locate the ends of the range of the nascent population, which by this time extended several miles beyond Chichagof Harbor. I spent the remaining few days walking the beaches in search of sea otter remains (I found none) and documenting the diet and activity of otters at the outer reaches of Chichagof Harbor. These preliminary data provided a distinct contrast with those I had obtained several years earlier from Amchitka, in that the otters of Attu fed almost exclusively on sea urchins and spent comparatively little time foraging and large amounts of time resting.

Although I counted tens of thousands of Steller sea lions and several thousand harbor seals on the ensuing survey of Attu aboard the *Aleutian Tern,* I saw no other sea otters. I departed Attu in late August, intent on mounting a long-term field program there in the years to come.

SIX

Return to Attu

THE GREATEST DANGER IN SCIENCE is falling in love with one's own ideas. Scientists are caught by this trap all the time. Most recognize the problem and are constantly searching for ways to put their ideas to critical tests. And so it has been with me.

Although I was reasonably sure that the striking differences in sea urchin and kelp abundances between the Rat Islands and Near Islands that John Palmisano and I had seen and reported in the early 1970s (Estes and Palmisano, 1974) were caused by sea otters, I also recognized that the evidence was far from definitive. These island groups were different places, and I worried initially that some other difference between them was responsible for their differing reef communities. I had no idea what that difference might be, but then one seldom does when on the wrong track. My principal challenge in the mid-1970s, as I prepared to work further on the sea otter–kelp forest system, was to put the otter–urchin–kelp *trophic cascade* hypothesis to a more rigorous test. The question I faced was how to do that.

In confronting this challenge, I turned to my graduate training in statistics and experimental design. Rigor in science requires experimentation, broadly defined as the application of some purported or hypothesized causal process (typically referred to as a "treatment") to the object of interest (which, depending on one's area of research, might be anything from subatomic particles to ecosystems) in such a way that any subsequent change can be inferred to have been caused by the treatment. The methods of experimental design and statistical inference were developed by Sir Ronald Fisher in the 1920s and '30s for the purpose of increasing rigor in pure and applied genetics and agricultural science. Fisher's methods have since been used in all areas of scientific research.

The power and appeal of a well-designed experiment is that it tests a scientific hypothesis in a way that excludes alternative hypotheses. The general idea is to begin with a number of similar experimental units. In genetic studies these experimental units are typically individual organisms; in agricultural studies they might be plots in a field. The treatment of interest is then applied to some of these units but not to others, so that the resulting difference between the treated and untreated units can be inferred to have been caused by the treatment.

Experimental design has three basic principles: replication, randomization, and local control. Replication, the repeated application of a treatment to otherwise similar experimental units, helps ensure that any result that was observed in the first run of the experiment is indeed recurrent and real. Randomization, the blind application of treatments to experimental units, assures that any resulting treatment effect is not caused by some other intrinsic difference among the experimental units. The process of randomization doesn't make these intrinsic differences among experimental units go away, but it does allow the experimenter to partition the overall variation in experimental results into two components: the treatment effect (that caused by the experimental treatment) and experimental error (the failure of identically treated experimental units to respond in exactly the same way to the treatment because of the intrinsic differences among them). Local control—the third basic principle of experimental design—endeavors to make the intrinsic differences among experimental units as small as possible, thereby reducing experimental error and making the treatment effect easier to see.

By the mid-1970s, the experimental method was finding its way into field ecology. Ecologists were beginning to recognize the limitations of simple descriptions, regardless of their detail or the degree of understanding of basic natural history. *Perturbation experiments*—studies in which a species suspected of playing important ecological roles was added to or taken away from some study plots and left unchanged in others—were revealing an otherwise invisible dynamic infrastructure of natural biological systems. People like Joe Connell and Bob Paine pioneered this approach in the 1960s with their studies of predation and competition in rocky intertidal communities. Others, like Tony Underwood, followed with reminders of how to conduct experiments using Fisher's principles of experimental design.

The experimental approach to ecological research, much of which was being developed in coastal marine systems and some of which involved sea urchins and kelp, helped me understand methodological strengths and weaknesses of my own work and what needed to be done to shore up the weaknesses. There were practical limitations to what I could do. Unlike the rocky intertidal systems that Connell and Paine studied, in which organisms like sea stars, whelks, mussels, and barnacles were moved around in specific and well-controlled ways, I couldn't purposely manipulate sea otter populations and was thus limited to what history and geography offered up. But as it was, history, geography, and unforeseen future changes in the distribution and abundance of sea otters across the Aleutian archipelago offered quite a lot.

My goal in the mid-1970s was to resolve whether differences in the abundances of sea urchins and kelp between the Rat Islands and Near Islands were truly caused by sea otters or by some other unknown and confounding influence. The answer, I thought, could be obtained from Attu Island, westernmost of the Aleutians and the largest in the Near Islands group. My survey of Attu in 1975 established that the small otter colony, which had reinvaded the island near Chichagof Harbor in the mid-1960s, was firmly established and growing rapidly. My plan, therefore, was simply to monitor the growth of this population over the years to come and measure whatever concurrent changes followed in the distribution and abundance of kelps and sea urchins. If the seascape of shallow reefs around Attu converged toward what I had seen in the Rat Islands, then the possibility of the differences between these island groups being caused by something other than sea otters could be discounted. Other islands at which otters remained either abundant or absent during the same period would serve as controls. I imagined that the *time series* of data from Attu would also reveal a functional relationship between otter abundance and kelp forest community structure.

So, I set out in 1976 to follow these time series in the years to come, realizing that I was committing myself to a long-term study and that for any number of reasons I might be derailed along the way. Replication of this mega-experiment was beyond my means and thus seemed out of the question, but I was willing to live with that.

Attu is literally at the end of the earth, so just getting there was no simple matter. I arranged with the Alaska Maritime National Wildlife Refuge to use their vessel, the *Aleutian Tern,* which operated in the area during summer months. Support from this vessel and its crew was essential in the beginning of my work on Attu because it provided a means of moving heavy materials and equipment to remote field sites. But traveling to Attu by ship would depend on the ship's schedule and area of operation. It also required a long boat ride—two days steaming from Adak and about five days from Dutch

Harbor. Over the longer term, I needed a means of transportation that was faster and could be more reliably planned. At the time, Reeve Aleutian Airways provided weekly commercial air service from Anchorage to Attu, mainly to fulfill the requirements of their mail-delivery contract with the U.S. Government. But because of Attu's mountainous terrain, commonly foggy weather, and the absence of ground-based electronics for aircraft approach and landing, the chance of actually making it to Attu on any scheduled flight day was about 50–50, and weeks sometimes passed between flights. The U.S. Coast Guard also flew C-130s from Kodiak to Attu every other week to supply the LORAN station. In contrast to Reeve, which had no reason to risk a landing on Attu during marginal weather conditions, the coast guard flights carried food and other important supplies, and therefore their crew would usually stand by at Shemya until conditions on Attu improved. So that was my best bet for reliable transportation to and from Attu.

An inquiry to Alaska Coast Guard Command in Juneau produced the hoped-for response. They were willing to transport people, supplies, and equipment on scheduled flights to Attu whenever space was available, which meant most of the time. I worked on Attu nearly every summer from 1976 to 1990, and during most of those years my field crew and I flew out and back from Kodiak with the coast guard. The drill became routine. We would show up on Kodiak a day or two before the scheduled flight to Attu, purchase food and supplies, and haul those things and our other gear to the coast guard's air station the night before the flight. Several months later, when the year's work was finished on Attu, we would hop a coast guard flight back to Kodiak, returning home via commercial air from there.

My first full field season on Attu in 1976 was challenging because none of the required field-support infrastructure was yet in place. I needed a small boat, a place to stay and store my gear, a means of transportation on land, an air compressor, and other various equipment required for diving. Mostly importantly, I needed a field assistant—someone I could count on to dive with me, safely run a small boat, fix things when they were broken, plan and manage logistics in my absence, and generally take care of him or herself in a remote and sometimes hostile environment. I had planned to hire Rich Glinski, who as an undergraduate at the University of Arizona had worked with me in the Aleutians during my graduate years. But after finishing his bachelor's degree, Rich was offered and accepted a better and more secure position with the Arizona Game and Fish Department. This apparent setback led me to David Irons (figure 6.1).

FIGURE 6.1. David Irons (left) with Elaine Rhode, Bob Mayer, and Charles Simenstad. (Photo by Charles A. Simenstad)

I met Dave in 1975 during a visit to Penn State University. Bob Anthony, a friend from graduate school and my faculty host at Penn State, had become acquainted with Dave through the university dive club. Bob knew I was looking for a field assistant and thought that Dave, who was about to graduate with a bachelor's degree, would be a promising candidate for the job. All I knew about Dave was that he liked to dive and that he had wilderness survival training. But Bob endorsed Dave as a man of substance and character, and that carried a lot of weight with me. The decision to hire Dave was a risk, but there were no better options so I took it.

My need for a small boat was easily resolved. The Atomic Energy Commission had left two 17-foot Boston Whalers at Amchitka when they departed the island in 1973, which had been turned over to the U.S. Fish and Wildlife Service. Since I now worked for the latter, I laid claim to one of them and had it sent over to Attu on the *Aleutian Tern*.

Lodging on Attu was a greater challenge. This was before the days of Weatherports, and a tent camp just wouldn't do for long periods of fieldwork, during which I knew we would frequently be wet and cold from diving and inclement weather. I needed a place to stay where we could warm up, dry out wet clothing, and take refuge during the region's frequent and sometimes intense storms. My plan for work on Attu involved activities at two sites: one at Chichagof Harbor on the island's north side, where otters were already established; and the other at Massacre Bay on the island's south side, which at the time was beyond the range of the expanding otter population. The landing strip and the coast guard's LORAN station were located in Massacre Bay. Some years earlier, following the transition from LORAN A to LORAN C, the coast guard station was moved from Casco Cove to a site near the landing strip, thus leaving the original LORAN A facility abandoned and unoccupied. These old buildings, dank and filthy with rat poop and urine, nonetheless were easily cleaned up and made habitable with an oil stove for heat and bunks and tables fashioned from scavenged lumber that had been left lying about after the war.

There were no such structures at Chichagof Harbor, so we had to build one. Dave designed the cabin, calculated the exact length of each piece of lumber needed, and cut and labeled the latter while in Anchorage. We did it that way because there was no electricity to run power tools at the field site, and making all the cuts with handsaws would have taken too much time. The precut lumber and other building materials were then shipped to Attu on the *Aleutian Tern* and deposited on the beach near our chosen campsite at the head of Chichagof Harbor. We fashioned a foundation from drift wood scavenged off a nearby beach and, with just hammers and a box of nails, assembled the cabin in several days. We added a small room for equipment storage the next summer. With a few minor modifications and repairs, my field crew and I lived in the cabin over the next 15 years.

Each spring the *Aleutian Tern* dropped off a 55-gallon drum of heating oil and two drums of gasoline at both sites, which, together with the food and other supplies we brought from Kodiak, was all we needed to support the fieldwork. Typically, we spent about half the summer working at one of the two sites, after which we loaded up the Boston Whaler with food and equipment and ran the 40 or so miles around the east end of Attu to the other site, where we worked for the remainder of the summer. It was a grand and exciting adventure, full of discovery and learning, in a spectacularly beautiful and wild place.

My reason for working on Attu was to chronicle the patterns of sea otter population growth and associated changes in the coastal ecosystem. I did this by establishing a series of standardized measurements, the plan being to repeat these measurements as the sea otter population grew to carrying capacity.

The study depended critically on measurements of the distribution and abundance of sea otters. I obtained these data by conducting surveys from the

Boston Whaler during calm, clear weather when the otters were easy to see and count. For each otter seen, I noted the location, group size, number of dependent pups, and whether the animals were within or outside the surface kelp canopy. These surveys were begun well before civilian availability of mobile GPS (global positioning system) technology, so I used a World War II navigation chart and assigned the location of each otter sighting to a series of contiguous coastal segments, each about 2–3 nautical miles in length and delineated by prominent landmarks, such as headlands or emergent offshore rocks. During any given year, I surveyed the coast of Attu in each direction from our camp at Chichagof Harbor until I found both ends of the otter population's range, which were easily recognizable by the presence of large groups of males. After locating these groups, I continued surveying for another 10 miles or so to ensure that all or most of the population's range had been counted. Each year, as the population grew and its range expanded, I had to travel farther and farther by skiff to survey the entire population. Eventually, sea otters spread around the entire island and a circumnavigation was required, at which point two days were needed to complete the survey. The close fit of the time series of counts to an exponential growth model (Estes, 1990) convinced me that the survey method was sufficiently accurate.

Often, either sea otters haul out on the shore to die or their carcasses wash ashore after death at sea. As a measure of natural mortality in the sea otter population of Attu, I searched two stretches of shoreline, each about 15 nautical miles in length, for fresh carcasses and older remains. One of these transects was near Chichagof Harbor and the other near Massacre Bay. In similar searches of the beaches at Amchitka Island during the early 1970s, I had found hundreds of sea otter carcasses, most of which were recently weaned juveniles or aged adults that had died as an apparent *density-dependent* response to competition for food in a population that was at or near carrying capacity. I anticipated finding few carcasses on Attu until the population reached carrying capacity, at which time I expected carcass recovery rates to increase.

I also expected diet and foraging behavior of the otters on Attu to change as the population grew toward carrying capacity and depleted their prey. In an effort to chronicle these expected changes, I planned to measure the diet and the *activity budget* of the sea otter population on Attu through time as it grew toward carrying capacity. The data were obtained from five locations along the northeast coast of Attu between Holtz and Sarana bays. Two kinds of information were obtained. First, I conducted focal follows of foraging otters (*focal animal sampling*), during which I recorded foraging dive times and the preceding and following surface times; whether or not dives resulted in a prey capture; and the number and species of captured prey. Second, I measured the activity budget of the population by slowly scanning the ocean's surface from one side of my observation post to the other and recording the activity (foraging, resting, grooming, interacting, or other) of each observed otter. I realized these data were biased against foraging activity because diving animals were more likely to be missed during a scan than those resting on the surface. Nonetheless, I reasoned that the bias should be constant through time and, thus, that any change in activity would be evident in the time series.

I expected the sea otter's diets to diversify and their time spent foraging to increase as the population grew toward carrying capacity. Additional dietary data were obtained from sea otter *scat* that I found along the shores and collected during carcass surveys. Although direct foraging observations are normally a much better means of measuring diet than searching through scat for prey remains, the scat proved useful in helping me determine the identity of certain prey species and, especially, in reconstructing the sizes of sea urchins that the otters had eaten. These latter determinations were possible because the urchin's calcified jaws, which are consumed by otters when they eat urchins, scale precisely in length with the urchin's test diameter and normally pass through to the scat undamaged. I was thus able to compare the sizes of urchins in the living population with the sizes that were eaten by otters (Estes and Duggins, 1995).

Most days on Attu during these early years were spent diving. I was interested in urchins and kelp, so my diving efforts were focused on documenting the distribution and abundance of these organisms. I measured the densities and the *percent cover* of kelps and other sessile organisms on Attu's subtidal reefs by placing a 0.5 square meter quadrat on the seafloor and either counting the individuals of each species (for density measurements of the larger organisms) or estimating their percent cover (for small or clonally living organisms). The quadrats were moved at randomly predetermined distances over the seafloor and resampled until 20 sets of measurements had been obtained. I sampled urchin populations in a similar way, except in this case I continued until 200 or more individuals had been obtained or 20 quadrats had been sampled, whichever occurred first. I measured urchin test diameters and used these data together with quadrat counts to characterize the population density and *population size structure* of sea urchins. The entire procedure was repeated each year at each of four sites, two near Chichagof Harbor and

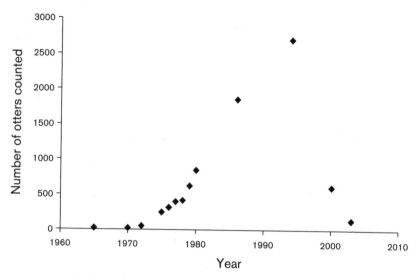

FIGURE 6.2. Numbers of sea otters counted around Attu Island from 1965 through 2003. The 1965 value is from R. D. Jones (1965).

two along the western margin of Massacre Bay. These samples, I thought, would provide representative measures of the abundance and species composition of kelps, the abundance and population structure of sea urchins, and whatever changes might occur in those metrics through time.

In those early years, I thought that the Attu time series might be all that I would have to test the otter–urchin–kelp trophic-cascade hypothesis. That expectation turned out to be wrong for two reasons. One is that the Attu sea otter population did not grow to carrying capacity. Judging from the amount of coastal habitat on Attu and otter population densities at Amchitka and Adak, I estimated an equilibrium population size of roughly 8,000–10,000 animals for Attu. The population was headed in this direction through the 1980s, but in about 1990 it suddenly began to decline rapidly, and by the early 2000s it contained fewer animals than when I began working there in 1975 (figure 6.2). The decline's cause became clear enough: the otters were being eaten by killer whales (Estes et al., 1998). While this stunning turn of events, which I will discuss further in chapter 10, was itself a fascinating ecological phenomenon, my reason for working on Attu suddenly vanished in the 1990s.

Although my vision for Attu dimmed with the sea otter's population collapse, tests of the otter–urchin–kelp trophic-cascade hypothesis would be replicated by the efforts of other people and through my own expanded studies in different parts of the sea otter's range. All of these various studies employed one or both of two general methods: contrasts between nearby areas with and without sea otters, and time series from particular areas as sea otter populations waxed or waned.

The first independent test was provided by David Duggins, a student from the University of Washington who worked in southeast Alaska during the mid- to late 1970s. Duggins's studies began in Torch Bay, a remote fjord on the exposed outer coast of Glacier Bay National Park. As in the Aleutians, sea otters had been overexploited in southeast Alaska during the maritime fur trade, although in southeast Alaska they were hunted to extinction.

In an effort to repatriate sea otters to southeast Alaska, in the late 1960s the Alaska Department of Fish and Game relocated 412 individuals from Amchitka Island and Prince William Sound to several locations, one of which was Surge Bay on the west coast of Yakobi Island (Jameson et al., 1982). Southeast Alaska is more than 1,500 miles from the western Aleutians and so, not unexpectedly, differences existed between these regions. For example, a single species of sea urchin, the green urchin (*Strongylocentrotus polyacanthus*), lives in the Aleutians; whereas two additional species, the red urchin (*S. franciscanus*) and the purple urchin (*S. purpuratus*) occur in southeast Alaska. The sunflower sea star (*Pycnopodia helianthoides*), an urchin predator, is common in southeast Alaska but rare or absent in the central and western Aleutians. There are other differences as well between these regions. But one feature of the subtidal reefs that Duggins encountered in Torch Bay in the mid-1970s was similar to what I had seen on Attu and Shemya: loads of urchins and relatively few kelps. Although Surge Bay is less than 30 nautical miles south of Torch Bay, the areas are separated by Cross Sound; thus, the growing Surge Bay sea otter colony had not yet spread northward across the sound and into Torch Bay when Duggins began working there in 1977.

Duggins tested the otter–urchin–kelp hypothesis in two ways. He first experimentally simulated the effect of otter predation by removing urchins from patches of reef habitat in Torch Bay while leaving other reef patches as unaltered controls. Kelp increased markedly in the urchin-removal plots but remained rare in the control plots, thus demonstrating that the overall lack of kelp in Torch Bay was caused by urchin grazing. This was no real surprise. Bob Paine and Bob Vadas had shown the same thing in 1969 by removing sea urchins from tide pools along the coast of Washington (Paine and Vadas, 1969). Even so, the confirmation of process was significant.

Duggins then went on to do what I had done in the western Aleutians—he compared Torch Bay, a site lacking otters, with Surge Bay, a site at which otters were abundant. The patterns he observed and reported were similar to what I had seen in the Aleutians. Sea urchins were rare and rocky reefs were covered with kelps in Surge Bay, whereas urchins abounded and kelps were rare in Torch Bay (Duggins, 1980). David's work was extremely important to me, in part because it was so well done and in part because it demonstrated the otter–urchin–kelp trophic cascade in a far-distant part of the North Pacific Ocean.

Evidence of widespread geographic occurrence of the otter–kelp trophic cascade mounted, with reports of similar differences between nearby sites with and without sea otters in British Columbia (Breen et al., 1982) and Washington (Kvitek et al., 1998). These comparative studies of areas with and without sea otters left little doubt in my mind about the existence of an otter-induced trophic cascade. But a nagging concern remained: places with and without sea otters inevitably also differed in ways that were not due to the effects of otters. That concern would never be laid to rest by such comparisons, regardless of what they showed or how numerous they might be. The point may seem trivial and silly, hardly worth the effort required to resolve it. But the psychological impact of such unresolved issues could be powerful in the minds of critical scientists who, for whatever reason, may not have wanted to believe our story or saw it as their job to point out its deficiencies. Furthermore, the data from such contrasts only provided a view of the end-points in a continuum of change between urchin barrens and kelp-dominated habitats. If sea otters were indeed responsible for reported differences between places with and without them, a transition between the two distinct community types would occur as the influence of these predators was added to or removed from the ecosystem. My desire to resolve uncertainties left by the comparative method, and especially to see and understand the functional relationship between sea otter numbers and kelp forest community structure, led to the second general method of testing the otter–urchin–kelp trophic-cascade hypothesis, which was simply to document patterns of change at particular locations as otter populations waxed or waned through time.

My work on Attu was the first attempt to obtain information of this nature. Even though the Attu sea otter population would collapse before I was able to fulfill the study's initial goals, I learned a lot about the functional relationship between sea otter abundance and reef community structure from what I saw and measured on Attu. The first hints of this functional relationship appeared in 1976, at which time I counted 318 otters around Chichagof Harbor and none in Massacre Bay. Although sea urchins were abundant at both sites, the largest urchins in Massacre Bay were more than 100 millimeters in test (skeletal) diameter, whereas the largest in Chichagof Harbor were less than 40 millimeters in test diameter. Since the urchin population structure in Massacre Bay was similar to what I had documented at Shemya, and the population structure in Chichagof Harbor was similar to what I had documented at Amchitka, I concluded that the difference between Chichagof Harbor and Massacre Bay was caused by sea otter predation.

Despite these differences in population structure, urchins were numerous in Chichagof Harbor and kelps remained rare. I couldn't determine whether kelps had increased in Chichagof Harbor from this early influence of sea otters, but if they had, it wasn't by much. Both areas were extensively deforested by sea urchin grazing.

I monitored Attu each summer through 1980, by which time the sea otter population had grown to 850 animals. Although otter numbers were increasing rapidly, densities and population structures of urchins and the distribution and abundance of kelps in Chichagof Harbor and Massacre Bay remained unchanged during this period. Moreover, the expanding sea otter population was still quite a long distance from my Massacre Bay study plots. By this time, I had come to realize that my hypothesized impact of otters on the Attu nearshore ecosystem was either nonlinear or incorrect, that more time was needed to resolve those alternatives, but that continued yearly sampling was not the best use of my time in the interim.

Other interesting questions had arisen during my initial five years on Attu, so I decided to cut back on monitoring—from yearly to every fourth or fifth year—in order to pursue these questions. I continued to dive in Chichagof Harbor and Massacre Bay each summer, thinking that if the system did begin to change, I would notice it and could resume monitoring. I had planned to resurvey Attu in 1984 or 1985, but the spreading population had not yet reached Massacre Bay in 1984; nor had the urchins or kelps changed appreciably in Chichagof Harbor or Massacre Bay since the onset of my sampling program in 1976. I waited. The first reinvading otters showed up near my Massacre Bay study sites in 1985; in 1986, I resurveyed the otter population and resampled the reef sites at Chichagof Harbor and Massacre Bay.

By 1986 the sea otter population had grown to 1,850 animals and had expanded its range along the south coast of Attu well beyond Massacre Bay.

The larger urchins suddenly disappeared from Massacre Bay, presumably eaten by the reinvading otters, and the urchins' population size structure became similar to that which I had seen from the study's onset in Chichagof Harbor. But there was no concurrent change in the distribution and abundance of kelp. Smaller urchins in Massacre Bay were as abundant as ever or more so, and these smaller urchins were preventing kelp recolonization. Like the reefs in Chichagof Harbor, Massacre Bay remained a largely deforested ecosystem, very different from what I had seen at Adak and Amchitka in the 1970s. These patterns persisted into the early 1990s, as the otter population of Attu continued to grow. I was beginning to wonder whether the reefs of Attu would ever shift from urchin barrens to kelp forests, thereby raising the possibility that the Near and Rat island reefs differed from one another in ways that were not attributable to sea otters. But I also knew that the Attu sea otter population was still less than a quarter of its estimated carrying capacity. Therefore, the grand experiment I had envisioned in the mid-1970s was far from having run its course, which under the most optimistic scenario (continued otter population growth at 17–20 percent per year) would require another decade.

But the Attu experiment would never run its course, because the sea otter population did not continue to grow toward carrying capacity, as I had expected through the 1970s and 1980s (figure 6.2). As the population trend abruptly turned from rapid growth to sharp decline in about 1990 and otter densities fell, the maximum size of sea urchins at Chichagof Harbor and Massacre Bay began to increase. As sea otter densities continued to decline through the late 1990s and early 2000s, the size of the largest urchins in both areas continued to increase. By 2009, when I last sampled Attu, the urchin population structures and overall reef communities at Chichagof Harbor and Massacre Bay were indistinguishable from those I had seen and measured at Shemya and Attu in the early 1970s, before otters had reinvaded these islands.

The collapse of sea otters in southwest Alaska was bad for the otters but good for what I was able to learn about them. The most important insights didn't come from Attu but from Adak and Amchitka, which supported sea otter populations at or near carrying capacity before the decline. The collapse of sea otters at Adak and Amchitka thus provided independent tests of the trophic-cascade hypothesis. More importantly, the data from Amchitka and Adak, together with the data I gathered from numerous other islands across the central and western Aleutian archipelago in the late-1990s through the end of the first decade of the twentieth century, defined a relationship between sea otter density and reef community structure for a declining sea otter population. The Attu time series from the mid-1970s through 1990 added to this by giving me a partial view of this same relationship for a growing otter population. Together these data sets greatly clarified my understanding of the functional relationship between otter density and kelp forest community structure.

In 1988, just before the onset of the sea otter collapse in southwest Alaska and when populations at Adak and Amchitka were still at or near carrying capacity, I surveyed the subtidal reefs at numerous randomly selected sites around both islands. As I'll explain further in the next chapter, I did this in response to suggestions that I had perhaps "cherry picked" sites in my earlier studies, thereby overextending the scale of impact from the otter–urchin–kelp trophic cascade. I discovered that kelps abounded and urchins were small and relatively rare at all sample sites at both Amchitka and Adak, thus resolving that issue. I assumed at the time that these islands had recovered from the maritime fur trade, were in a steady-state equilibrium, and therefore would not change appreciably in the foreseeable future. There was no compelling reason to resample the sites at Amchitka and Adak.

At the time, I saw little opportunity or need for further testing the otter–urchin–kelp trophic-cascade hypothesis in the Aleutians, and my work there turned toward the otters themselves and trying to understand the mechanisms of population regulation at carrying capacity. I had occasion to dive at Amchitka again in 1994, but I didn't resample the reefs, which appeared to be largely unchanged except for a modest increase in urchin abundance (Watt et al., 2000).

I initially attributed the 1994 increase in urchin abundance at Amchitka to the inherent messiness of field data. That interpretation was incorrect. By 1994, the sea otter populations at Amchitka and Adak had already declined by about 50 percent, and the urchin increase turned out to be the beginnings of a response to the otter's decline. But I hadn't yet come to grips with the possibility that otter numbers were declining, despite what in retrospect were many indications, including the results of a 1994 otter survey at Amchitka, which produced a count that was only about half of what I thought it should be. I attributed all of this to the combined effects of sampling variation and the "natural" fluctuations that characterize all populations. The problem was in my mind-set. I thought that the otter populations and their ecosystems should be stationary, so that's the context in which I interpreted everything

I saw. It wasn't until several years later, by which time I had left Amchitka and was again working at Adak, that I finally accepted the reality of the decline.

Once this reality had sunk in, I resurveyed the subtidal reefs at Adak (in 1997) and was stunned by what I saw. Urchins were suddenly everywhere, and they were mowing down what remained of Adak's kelp forests. All of Adak's more than 30 reef survey sites, which supported healthy kelp forests when I sampled them in 1988, had become sea urchin barrens. I resurveyed Amchitka in 1999 and discovered that the kelp forests had transitioned to urchin barrens there as well.

Once I realized that otter populations were collapsing across the entire Aleutian archipelago, I thought it would be important to monitor the otters and their associated ecosystems from as many places as possible. I did this through 2010, by which time otter densities had fallen to less than 5 percent of their estimated carrying capacity at every island, and the associated reef communities were indistinguishable from those that lacked otters entirely. These data eliminated any uncertainty I may have continued to harbor over the strength of evidence for the otter–urchin–kelp trophic cascade. But more importantly, I now had a large data set with which I could relate kelp forest community structure to otter population density over its full range of possible variation—from zero to carrying capacity.

A closer look at this information revealed several interesting patterns. The reef communities always occurred in one or the other of two states—as kelp forests or sea urchin barrens (figure 6.3). There were no intermediate states, even though associated sea otter densities took on the full range of intermediate values. These collective data confirmed what I had begun to suspect—that the transitions between kelp forests and urchin barrens occurred as abrupt *phase shifts* (discussed below). The *phase states* in these Aleutian systems were strongly related to otter densities. However, the functional relationship between otter density and phase state differed markedly, depending on whether otter populations were growing or declining. Systems that began as urchin barrens with the reestablishment and growth of sea otters remained in the urchin-dominated state until otter densities approached carrying capacity. Conversely, systems that began as kelp forests, with otter populations at or near carrying capacity, remained in the kelp-dominated state until otter densities had fallen by more than half (figure 6.4).

The expansion of sea otter populations in southeast Alaska and British Columbia provided similar tests of the otter–urchin–kelp trophic-cascade hypothesis. When David Duggins first surveyed Torch Bay in the late 1970s,

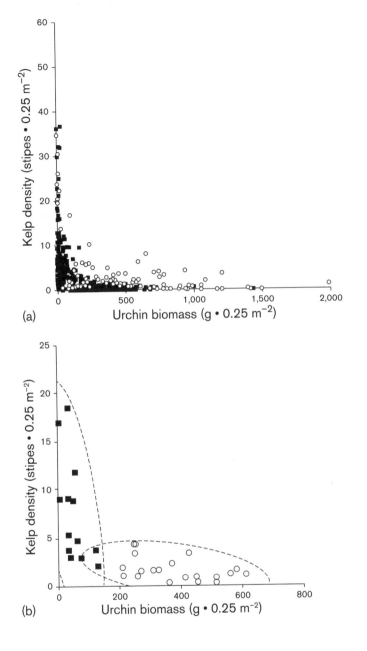

FIGURE 6.3. (a) Kelp density in relation to urchin biomass in the Aleutian archipelago, measured in 0.25 square meter quadrats at 463 sites at 19 islands between 1987 and 2006. (b) The same data, averaged by island–time combinations. Open circles represent sea otter densities of fewer than six individuals, and filled squares densities of more than six individuals, per square kilometer of shoreline. Dashed lines are 90% confidence ellipses around the two aggregates identified by K-means cluster analysis, which reflect the kelp-dominated (filled squares) and urchin-dominated (open circles) phase states. The graphs are from Estes et al. (2010b).

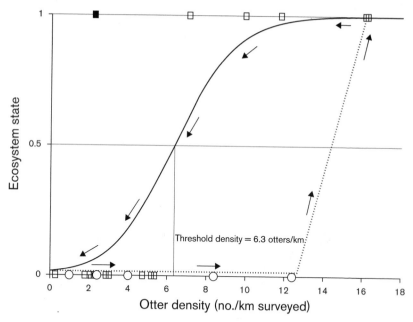

FIGURE 6.4. Trajectories of reef phase states with increasing (dotted line) and declining (solid line) sea otter populations in the central and western Aleutian Islands. Ecosystem state is either deforested (urchin barren, o) or kelp forest (1). See Estes et al. (2010b) for further analytical details.

otters were absent and the reefs were extensively deforested. Sea otters reinvaded Torch Bay in the mid-1980s, so David and I decided to resurvey Torch and Surge bays in 1988, wondering to what extent the areas had changed over the preceding decade. Our only surprise was in the magnitude of the change. Kelps now covered the Torch Bay reefs, with scarcely an urchin to be found. The Surge Bay reefs were kelp dominated, as they had been in the late 1970s. My colleagues and I have resurveyed both areas several times over the ensuing 25 years, during which period they have remained kelp dominated.

In 1988, expanding sea otter populations in southeast Alaska had not yet spread into Sitka Sound, which is about 65 nautical miles south of Surge Bay. So later that summer, with the help of Diane Carney (a former graduate student) and Glenn VanBlaricom (a former colleague), I surveyed the reefs at 30 sites in Sitka Sound, my intent being to use the data to characterize another otter-free location and to document whatever changes occurred after otters spread into Sitka Sound in the years to come. Every site in Sitka Sound was covered with red and purple sea urchins. Not one individual kelp plant turned up in the hundreds of sample plots, and few were found by more extensive looking. Jim Bodkin (a sea otter biologist with USGS in Alaska), George Esslinger (Bodkin's coworker), Keith Miles (a colleague from California), and I resurveyed these same sites in 2009, two or three years after sea otters had expanded into Sitka Sound, and saw the astonishing but now familiar pattern of change. The reefs all had become kelp covered, and there were hardly any urchins.

Between 1969 and 1972, Canadian and Alaskan resource agencies reintroduced 89 sea otters in Checleset Bay, on the northwest coast of British Columbia's Vancouver Island. Jane Watson, a former student of mine, used this expanding otter colony to study associated changes in the reef community (Watson and Estes, 2011). She did this by establishing 20 sample sites at each of three locations: Checleset Bay, Kyuquot Sound (about 20 kilometers south of Checleset Bay), and Barkley Sound (about 400 kilometers farther south from Kyuquot Sound). She then repeatedly sampled each of these sites over the next 20 years, during which time otters remained present in Checleset Bay, reinvaded Kyuquot Sound, and remained absent from Barkley Sound. Kelps were abundant and urchins rare throughout the 20-year period in Checleset Bay. Likewise, urchins were abundant and kelps rare throughout the period in Barkley Sound. In Kyuquot Sound, the sites transitioned from urchin barrens to kelp forests within months after the otters arrived.

The collective findings from more than 40 years of field research at numerous sites across the Pacific Rim provide unequivocal evidence of the otter–urchin–kelp trophic cascade. The widespread occurrence of this ecological process is indisputable, at least to my mind. The more interesting insight has to do with the time course of change, which differs markedly with otter densities, depending on region and whether sea otter populations are growing or declining. Changing otter densities always lead to abrupt shifts between kelp-dominated and urchin-dominated reef states. As a rule of thumb, systems that are urchin dominated tend to remain urchin dominated, and systems that are kelp dominated tend to remain in that state as well. This accounts for the differing trajectories of phase-state change with growing and declining sea otter populations. As another rule of thumb, the phase-state transitions occur at high otter densities in the Aleutians and at comparatively low otter densities in southeast Alaska and British Columbia.

To put all this in terms of conventional ecological jargon, the otter–kelp forest system is characterized by phase shifts (i.e., the changes are rapid and

FIGURE 6.5. Brenda Konar.

dramatic), *alternative stable states* (once a system falls into a particular phase state, it tends to remain there), and *hysteresis* (the functional response of the system to a controlling force—sea otters in this case—differs depending on whether that force is being added or taken away).

Why do reef communities change so abruptly between the kelp- and urchin-dominated states; why do the trajectories of change in response to otter abundance differ, depending on whether otter numbers are growing or declining; and why do the overall time-series patterns differ so strikingly between the Aleutians and the continental mainland of North America? We don't have all the answers. But we do have some of them.

A few of these answers came from the work of my former graduate student Brenda Konar (figure 6.5), who did most of her dissertation research at Shemya in the Near Islands. Although the reefs around Shemya were largely deforested when Brenda was there in the mid-1990s, patches of kelp occurred here and there. Brenda noticed that the spatial transitions between kelp patches and surrounding urchin barrens were abrupt and reasoned that these transitions could serve as an experimental model for understanding the similarly abrupt temporal transitions between the kelp- and urchin-dominated phase states that I and others had seen and reported earlier. Her reasoning was brilliant in its logic, its simplicity, and the fact that it allowed her to put various hypotheses to direct experimental tests.

Brenda's first effort was to describe the spatial distribution of the kelp and urchin phase states in detail. Besides confirming that urchin and kelp densities changed abruptly along transects across the transition zone, she discovered that urchins in the barrens were highly mobile whereas those in the kelp patches were sedentary. She further discovered that the gonads of urchins living in kelp patches were large whereas those in the adjacent urchin barrens were small. These findings suggested that urchins in the kelp beds were well fed by detrital fallout from the overlying kelp forests, whereas those in the adjacent barrens were undernourished and constantly on the move in search of food.

Brenda's early findings explained why the kelp patches didn't invade the urchin barrens, but they didn't explain why urchins from the barrens failed to invade the kelp patches. She discovered that if she experimentally removed the kelp from a small area along the transition zone, urchins would invade and then hold that cleared area in perpetuity. She expanded these experiments by moving urchins into plots at the center of the kelp patches, some of which had the kelps removed and others of which did not, with the same result. That is, the urchins remained when kelp was removed but disappeared rapidly when it was not. Brenda's finding that urchins inhibited kelp was no surprise. But her finding that kelps, once established, were capable of inhibiting urchins *was* surprising. *Context dependency* and *priority effects* clearly influenced the manner in which urchins and kelp interacted with one another.

Brenda's work changed my view of the way in which urchins and kelps interact and helped me understand the dynamics of time change between the kelp- and urchin-dominated phase states. Urchins often eat kelp by way of a typical consumer–prey interaction, in which the consumer (urchins) benefit and the prey (kelp) incur a cost. But it doesn't always work that way. The kelps, through their supple morphology together with the constant wave-induced movements of the surrounding water, are able to fight back invading urchins by a whiplash motion. This explains why urchin barrens, once established, tend to remain as urchin barrens, and why kelp forests, once established, tend to remain as kelp forests. Pushed far enough by a massive front of invading sea urchins, the kelp defense mechanisms break down. Pushed far enough by increased detrital kelp input, the urchins stop moving, so that they no longer have a destructive effect on the living kelp. These processes explain why the kelp–urchin phase shift displays hysteresis.

Although Brenda's experiments helped me understand the dynamic processes underlying the kelp forest phase shifts, the existence of alternative stable states, and why the functional relation between otter abundance and phase state is characterized by hysteresis, they did not explain why the transitions occurred at such vastly different sea otter densities between the Aleutians and southeast Alaska–British Columbia. I'm still not entirely sure why that is, but I've come to suspect that the difference arises from regional variation in physical oceanography and geology.

Adult sea urchins spawn regularly across the North Pacific, year after year. Male and female gametes join together in the water column to create *planktivorous* larvae. These larvae remain in the water column for a month or more, during which time they develop, grow, and are moved about by ocean currents. During most years these larvae are vectored away from the coastal zone of western North America by *Ekman transport,* the seaward movement of surface waters from the bending influence of the *Coriolis effect.* Only during some rare *El Niño–Southern Oscillation* years, when Ekman transport weakens or reverses at just the right time, do urchin larvae find their way back to shallow coastal reefs to settle. These processes and the resulting pattern are fairly well known. As a consequence, urchin populations along the western continental margin of North America are characterized by an absence of small individuals. Since foraging otters are strongly size-selective predators on the larger urchins, a front of reinvading otters will quickly drive urchin populations to very low numbers.

While the above-described patterns and processes characterize coastal environments of southeast Alaska and British Columbia, details of the physical oceanography and larval ecology of urchins in the Aleutians are poorly known. However, one clear difference from southeast Alaska and British Columbia is that urchins recruit heavily and predictably in the Aleutians. I can surmise this from the structure of urchin populations I have sampled at many places over many years, nearly all of which include large numbers of baby and smaller urchins.

I know that this difference in the frequency and strength of urchin recruitment between the Aleutians and the continental margin of western North America is not a consequence of regional differences in spawning. Therefore, it must result from differing fates of the larval urchins after the adults spawn. I suspect that one key process is the absence of offshore Ekman transport in the Aleutians, mostly because variation in this physical process is the cause of variation in larval settlement along the continental margin. Regardless of the cause, these smaller urchins—which are nutritionally worthless to, and thus not consumed by, the foraging otters—are nonetheless capable of preventing young kelp sporophytes from settling and becoming established. The Aleutian system thus tends to remain in the urchin-barren phase state, even in the presence of fairly high otter densities. Although I've never seen it happen, I suspect that the otters do eventually begin eating these small urchins as otter populations approach carrying capacity and starvation-induced hunger presses them to the point of consuming anything they can find. This general view explains why I never saw a shift to the kelp phase state on the reefs of Attu Island.

The spectacular collapse of kelp forests at Adak and Amchitka islands was accompanied by another notable event: the rapid appearance of large urchins. I was perplexed by the size of these animals because I knew that urchins in the Aleutians grew slowly and I couldn't figure out how they were becoming large so quickly. For years I had suspected the urchin populations in the Aleutians of being subsidized by deep-water immigrants, because I couldn't reconcile the urchin population's sustainability with their density, size structure, and individual growth on one hand and the numbers that were being eaten by otters on the other. I knew that otters almost never dove to depths beyond about 10 meters to feed on urchins. I also knew that large urchins occurred in deeper water (100 meters), because I had collected them from those depths. I had thus grown to suspect a sort of conveyor belt of urchins from deep water, where many of these animals settled and grew early in life, to the shallow zone where growth and reproduction would benefit from elevated production (but where the risk of being eaten by an otter was also much higher).

The conveyer-belt hypothesis explains rapid increases in urchin density and size that followed the collapse of otters at Adak and Amchitka. However, it also begs the question of why deep-water urchins were not pressing into the otter-dominated reefs of southeast Alaska and British Columbia. The explanation may lie in the fact that the Aleutians are oceanic islands whereas the coastal reefs of southeast Alaska and British Columbia are separated from deep water by a wide continental shelf. Urchins in the Aleutians might easily follow the steep and predominantly rocky substrate from deep water to shallow coastal reefs. In southeast Alaska and British Columbia, deep-water urchins would have to traverse a broad and often sandy or muddy continental shelf to do the same thing. I've tried to test this idea by looking for deep-water isotopic signatures in the larger shallow water urchins, but so far the data are equivocal and therefore the jury is still out.

I set out in the mid-1970s to more rigorously evaluate the otter–urchin–kelp trophic-cascade hypothesis, which at that time was based entirely on differences between two islands: Amchitka, where otters abounded; and Shemya, where otters were absent. The advantage of such "space for time" approaches in ecology is that they can be executed quickly; the disadvantage is that they cannot account for the potential confounding effects of other differences between the different places. My early vision was to resolve this uncertainty by following the changes at one particular place—Attu Island—through time as sea otters became established and grew toward carrying capacity.

All such long-term visions are subject to high risks of failure, and mine failed with the unexpected collapse of sea otters across southwest Alaska in the 1990s. But my failed vision turned out to be a blessing in disguise, because the collapse of sea otters at Amchitka and Adak islands provided a more powerful, compelling, and informative time series of data than I ever imagined would come from Attu. Moreover, I wasn't alone in my efforts to understand the otter–urchin–kelp trophic cascade through the use of "space for time" contrasts and time-series change. These efforts, several of which I joined as they were developing, provided a body of evidence in support of the otter–urchin–kelp hypothesis that was both unequivocal and extended the geographic scope of the phenomenon to a vast area of the North Pacific Ocean. The collective body of work also provided intriguing insights into the functional relationships between sea otter density and kelp forest community structure: their fundamentally nonlinear nature; differing trajectories between systems in which the otters were increasing or declining; striking geographic differences in otter densities needed to push the urchin–kelp phase shift across its breakpoint; and the rich tapestry of natural history underlying all of this generality and variation.

If there is a lesson here to aspiring young ecologists, it is this: Build your careers on a vision, but be prepared to take advantage of that which you will never anticipate in the beginning.

SEVEN

Generality and Variation

THE SEA OTTER–KELP TROPHIC CASCADE was becoming well known by the early 1980s. My work in the Aleutians, together with studies by David Duggins from southeast Alaska (Duggins, 1980) and Paul Breen and colleagues from British Columbia (Breen et al., 1982), portrayed the coastal North Pacific Ocean as an ecosystem that varied radically and predictably with the simple presence or absence of sea otters. But kelp forest ecologists elsewhere remained skeptical. Their doubts were not so much over the trophic cascade's existence in nature, but rather over its relative importance compared to other ecological processes that influenced the distribution and abundance of sea urchins and kelp. These doubts were fueled by two main observations: the occurrence of large kelp forests at many locations without sea otters in Southern California and Mexico, and strong effects of the El Niño–Southern Oscillation on the distribution and abundance of kelp across this same region. During one particularly strong El Niño event in 1978–79, kelp forests in Southern California declined drastically, and those along the west coast of Baja California essentially disappeared.

Michael Foster (a professor at Moss Landing Marine Laboratories) and David Schiel (his post doc at the time) were the most outspoken critics. In an effort to put the otter–kelp paradigm to a critical test, Foster and Schiel conducted a survey of research that had been done on subtidal reefs in California by as many people and at as many places as they were able to identify. Some of the work had been done within the sea otter's roughly 250-mile range in central California and some had been done beyond the otter's range, in Northern or Southern California. Foster and Schiel treated this information as a large-scale sample of sites with and without sea otters. In some cases they queried the individual researchers about the degree to which their study sites were kelp dominated or urchin dominated. In other cases they made these determinations themselves from contributed data or published findings. On the basis of these data and analyses, Foster and Schiel (1988) wrote a book chapter titled "Sea otters and kelp forests: keystone species or just another brick in the wall?" Their main conclusion, which is easily inferred from their paper's provocative title, was that the sea otter–kelp forest trophic cascade is just one of many factors influencing the structure and function of kelp forest ecosystems. This article posed a direct challenge to the application of our ecological model in California and, by implication, elsewhere.

I was beginning to realize that the ecological story about sea otters and kelp forests that my colleagues and I had assembled and published for the Aleutian Islands, southeast Alaska, and British Columbia was an oversimplification if applied to how kelp forest ecosystems worked in California. On the other hand, I was skeptical of Foster and Schiel's analyses and conclusions because I suspected that most people who studied kelp forests sought out kelp canopies as the locations to dive and do their research, thus almost certainly biasing the data toward the presence of kelp and away from the existence of sea urchin barrens. For example, one of the data sets in Foster and Schiel's analysis came from a survey done by Jack Engle and his associates at San Nicolas Island in Southern California. I had participated in one of Jack's surveys of nearby San Clemente Island and knew from this experience that they operated in just the way I suspected many of the other surveys were done—by first seeking out a surface kelp canopy and then surveying the surrounding habitats. Not surprisingly, Jack's data indicated that the reefs surrounding San Nicolas Island were more than 95 percent kelp dominated, with urchin barrens being a rarity. I had dived extensively around San Nicolas and knew that many of the reefs around the island were urchin barrens. As a result of the apparent discrepancy between Foster and Schiel's portrayal of Jack's survey data and my own observations at San Nicolas Island, my group surveyed 35 randomly selected sites around the island. This latter survey indicated that about half the area was deforested by urchin grazing and the remaining half was in the kelp-dominated state, a finding that differed not only from Foster and Schiel's characterization of the island, but from my own conceptualization of more northerly sites with and without sea otters.

Although I doubted Foster and Schiel's results, I recognized that tough new questions about the sea otter–kelp trophic cascade in more northerly ecosystems were raised by their critique, by my own data from San Nicolas Island, and by the obvious complexity of Southern California kelp forests (as revealed by the large number of studies than had been done there by various independent workers). I knew that the trophic cascade occurred in that region, but I also wondered if I had overstated the importance of this process in relation to other factors that influenced the distribution and abundance of sea urchins and kelp. Privately, I worried that I might have been guilty of the same procedural error I had accused Foster and Schiel of making—that is, working in places that were most likely to provide the kinds of results I was looking for. While I had seen enough of the central and western Aleutian Islands to be reasonably sure that my reported findings were robust, I also realized that my data were not obtained in such a way that would convince others of that. Something more was needed to address the question of generality and variation in the sea otter–kelp forest trophic cascade.

Both the need and the opportunity for resolving this question of generality and variation surfaced in the mid-1980s with a National Science Foundation–funded grant to explore the influences of kelp production on coastal food webs in the Aleutian archipelago. The proposal for this work was initiated by David Duggins and Charles (Si) Simenstand at the University of Washington. David and Si knew that kelp forests were highly productive and wondered how this production might be influencing coastal ecosystems in the North Pacific Ocean. They were drawn to the Aleutians as a place to answer that question because islands with and without sea otters also differed greatly in kelp abundance. I became involved because of my knowledge of the region and of sea otters in particular.

The premises underlying our proposal to the National Science Foundation were simple. We knew that islands with and without sea otters differed in kelp abundance; we suspected kelp to be important in the fueling of *secondary production;* and thus we hypothesized that islands with and without otters would differ in both the rate of secondary production and the relative contribution of kelp-derived organic carbon to the myriad species of animals in coastal ecosystems.

The testing of this hypothesis forced me to confront a more complex model of ecological process than I had considered in the past. In contrast to living kelps, which are sessile organisms affixed to an exact spot on the seafloor, kelp-derived carbon moves around as detritus (particulate organic carbon, or POC) and dissolved organic carbon (DOC) in the flow of surrounding seawater generated by wave surge and coastal currents. This potential for kelp carbon transport created a need to know both the extent to which kelp-derived organic carbon is moving around in nature and the degree to which the otter–kelp trophic cascade is generalizable at scales that were larger than those I had previously worked at.

As I contemplated these issues and the challenges they presented, I once again turned to the foundations of statistics, in this case to the principles of sampling and statistical inference. The purpose and goal of sampling, as almost everyone knows, is to accurately estimate some feature of a population. "Accuracy" and "population" are the key terms in any such endeavor. Exactly what do these words mean? The accuracy of any sample-based estimate is further divisible into two elements—bias and precision. A biased estimate is one that varies consistently in one direction or the other from the true value. Precision relates to the repeatability of the measure from sample to sample. In the best of all worlds, one would like for an estimate to be unbiased and precise. But bias and precision do not go hand-in-hand. A precise estimate can be deeply biased, and an unbiased estimate can be highly imprecise. Precision is easy to measure but often difficult to do anything about. Bias is more difficult to estimate, but once known it is easy to correct for. I knew that reef ecosystems varied considerably from place to place and, thus, that any estimate based on samples across these places would be inherently imprecise. There was nothing I could or should do about that. Instead, I was seeking an estimate of the sea otter's influence on sea urchins and kelp that was unbiased.

Statistical inferences are commonly conjectures about populations that are based on samples taken from those populations. The legitimacy of the inference depends on a clear and unambiguous definition of the population. But this fundamental aspect of statistical inference has been a tricky business in ecology, in part because biologists and statisticians define populations in different ways. Biologists think of populations as groups of broadly co-occurring and interacting individuals belonging to the same species. Statisticians, by contrast, think about a population as the "sample space." The sample space is formally defined as all possible outcomes of an experiment, where an experiment is defined as anything that generates a result. A statistical population, therefore, can be many kinds of things. For example, a statistical population might be the number of individual organisms in some area (the traditional biological definition); it might be the number of stars in all the galaxies of the universe; it might be the length of performance of a particular kind of lightbulb; it might be the distance driven to and from work by people in the state of California; or it might refer to the abundances of all of the component species of some ecosystem over some prescribed dimensions of space and time. A statistical population might also refer to the outcome of a species interaction over these same dimensions of space and time or, more explicitly, to the results of a manipulative experiment designed to measure the outcome of that interaction if the experiment were repeated ad infinitum. In the case of ecological populations, boundaries are typically arbitrary and commonly undefined, largely because the abundances of individuals are so variable and the distributions of species in space and time almost never coincide. In this regard, ecologists are almost never explicit about what they are measuring or the exact hypotheses they are testing.

Despite my doubts about Foster and Schiel's book chapter, the questions it raised and the needs of our kelp production project in the Aleutians forced me to think about these things. In particular, they forced me to redefine my immediate goals on the basis of four specific questions:

1. What do I need to measure?

2. How do I sample the measured entities at any particular site?

3. How can I array the sites in a way that provides a representative (unbiased) portrayal of the sea otter's influence over larger dimensions of space and time?

4. How can I define this sample space so that the "population" about which I am trying to make an inference is as explicit as possible?

The question of what to measure was easily answered. I would measure what I had measured before: the abundance of kelp and the density and size (test diameter) of sea urchins. I added a few more things to the sampling protocol, including the relative ages of the kelps (recruits vs. adults) and the percent cover of clonal or encrusting algae and invertebrates, mostly because these organisms were common and easily measured and I thought that the data might later be of value. How to sample these entities was also easily resolved in a satisfactory manner. I would simply drive a small boat to the sample site (defined by coastal line-ups or, later, by GPS coordinates), anchor at the desired depth, and begin sampling from the anchor by swimming a predetermined but randomly selected number of kicks along the depth contour, at which point a quadrat (0.25 square meter) was placed on the seafloor, within which the various measurements and samples would be taken (this procedure was similar to the one I had used in my earlier work in the Aleutians).

How to select locations to sample was a question that had to be resolved at two spatial scales: at the level of islands, and at the level of sites within islands. Inasmuch as we would be traveling from island to island by ship across the Aleutians and I already knew which of these islands either lacked otters or supported dense populations, the choice of which islands to sample was mostly imposed upon me by the sea otters.

Our cruise would start from Attu and progress eastward to Dutch Harbor on Unalaska Island in the eastern Aleutians. We would have a day on Attu to work out the kinks of the sampling protocol and then proceed eastward about 35 miles to the Semichi Islands, where sea otters were still absent. I was less interested in the data from Attu because at that time the island was in the early stages of recolonization by sea otters. We would begin in earnest by sampling Alaid, Nizki, and Shemya islands (otters absent), then move on to Amchitka and Adak islands, where otters abounded in populations that were at or near carrying capacity. Allowing for weather and steaming time, this would consume most of our two-week cruise.

I had more latitude in the selection of sites within islands and thus decided on the following procedure to ensure that the data obtained from any particular island were representative of that island as a whole. First, I placed a square grid over a map of the island and marked each gridline's intersection with the shoreline as a potential sample site. From this array of potential sites, I randomly selected about 35 to sample; I had no idea at the time what the minimum or optimum number of sample sites might be. The number 35 was based solely on logistical constraints: the number of dive teams I would have during the research cruise, the number of sites each team could sample in a day, and the number of days we would spend at each island. Thirty-five seemed like a large number, but I would have to wait and see if the resulting data were sufficiently precise to reveal patterns. I thought they would be—but I also worried that the data would be a mess.

Once in the field, I had to modify the sampling procedure for logistical reasons. Some of the islands were too large to sample in their entirety; for those, I chose the sample sites from a stretch of shoreline that could be safely reached from our base of operations, depending on the weather and sea conditions at the time of our visit. Except during unusual periods of calm weather, when we could dive anywhere, we would be forced to work on the leeward (sheltered) sides of the various islands. This latter constraint didn't worry me because there is no prevailing wind direction in the Aleutians and so the leeward sides of islands on any given day were also essentially random. As a result, some of our samples were from the Bering Sea (the northern coastlines), some were from the Pacific Ocean (the southern coastlines), and some were from the passes between islands. The areas within which I selected the sample sites were defined by about 20 linear nautical miles of coastline. Although this excluded much of the coastline of the larger islands from sampling, the sample areas were selected arbitrarily, depending on weather conditions and with no prior knowledge of their reef communities. Because of this, I assumed that our sample areas fairly represented the island as a whole. But even if that were not the case (and I would never know if it was or wasn't), any more local island effects ought to be subsumed in the replication of sample islands.

Defining the sample space for this very large-scale analysis was more problematic. To what exact geographic region would the inferences I might make from these data apply? Beginning with this cruise in 1987, we defined the sample space as the region circumscribed by Attu in the west and Adak in the east: a straight-line distance of about 425 nautical miles that included several thousand miles of coastline. Although this seemed like a large area over which to attempt a rigorous description of ecological pattern, in fact it was only a small part of the geographic range—less than 10 percent of the roughly 6,000 straight-line miles that define the region (from Hokkaido to southern Baja California) over which sea otters and kelp forest once occurred in the North Pacific Ocean.

Following the cruise I was acutely aware, on one hand, of how much we had accomplished—and, on the other, of how little it really meant. I would thus expand the effort over the next 20 years with sampling programs across the eastern Aleutian Islands, along the Alaska Peninsula, in southeast Alaska, and along the coast of British Columbia. The sample space for which we can infer the strength and nature of the otter–kelp trophic cascade thus now extends from the Commander Islands in the west to southern Vancouver Island in the southeast, a coastal straight-line distance of about 3,000 nautical miles. My accumulated database represents the center of the sea otter's geographic range (from the Commander Islands to southern British Columbia) but not its extremes (Kamchatka through the Kuril Islands and the state of Washington through Baja California). Ecologists know a lot about the geographic distributions of species, but very little about the distributions of species interactions. These data arguably provide the most expansive assessment of a species interaction that has ever been undertaken. What did we learn from the effort? In the remainder of this chapter, I'll recount the fieldwork and highlight the key findings.

Our initial survey of the western and central Aleutians was done from the R/V *Alpha Helix* (figure 7.1), which we were to meet up with and board from

FIGURE 7.1. The R/V *Alpha Helix* at Segula Island in the Rat Islands group. (Photo by Norman S. Smith)

Attu in late July 1987. The day we embarked, Attu was about as calm as I'd ever seen it, with virtually no wind and a barely perceptible swell. I was relieved to know that diving conditions would at least be good in the beginning. But the *Alpha Helix* was a roller if ever there was one, and I was amazed to see her rolling some 15–20 degrees as she steamed into Massacre Bay and set anchor in the nearly flat, calm ocean. I wondered how it would be to work from this ship when the weather inevitably worsened, and I worried, in particular, about the safety of moving skiffs and divers back and forth between the ship and the surrounding ocean. These concerns turned out to be unwarranted. I would use the *Alpha Helix* extensively in the years to come. She always rolled, and we just got used to it.

We boarded the ship about midday and began diving right after lunch. Because of the long summer days and the extreme westward extension of the Aleutian time zone, we were able to continue working until after midnight that first day, by which time we had managed to sample eight sites with just two dive teams. I was encouraged by this, realizing that three dive teams should be able to sample 30 or more sites in two full days. Later, as we became familiar with the routine, our efficiency increased and we were able to accomplish that goal.

We departed Attu for the Semichi Islands at about 1:00 A.M., expecting to be anchored and ready to work before dawn after the three- to four-hour overnight steam. But the weather "went south" overnight, and by morning it had become windy and the sea too rough to dive or to put a skiff in the water. We spent the day anchored in the shelter of Nizki Island, with our captain's assurance that the storm would be short lived.

The weather was much improved the following morning. As we prepared to begin diving and sampling in the Semichi Islands, I wondered (and secretly worried about) what the data we were gathering would ultimately tell us. Were Foster and Schiel right? Was living nature simply too variable for us to draw any broad inferences about pattern and process? Or would the sea otter–kelp trophic cascade hold up over large scales of space and time? I hoped that it would, of course, because otherwise the foundation of our current research program on kelp production would be seriously compromised.

The first site I sampled was near the northwest end of Alaid Island, adjacent to a dramatic cliff that dropped into the sea from a rise known as "Alaid Head" (the Semichi Islands are otherwise quite flat). It was early, and although the water was very clear, I couldn't see to the seafloor through the surface glare. But as I slipped over the skiff's side into the water, I could clearly see the seascape below. At this site, at least, it was just as I had expected—urchins everywhere and barely a kelp plant to be seen. As my dive partner and I completed the sampling, I knew what the data from this site would show.

We moved on to the next randomly selected site and it was more of the same—lots of urchins and little or no kelp. By late morning, when we returned to the ship for lunch, we had sampled four sites, all of which were similar. I queried the other two dive teams, both of which had sampled four sites but in different areas. They reported much the same. We now had sampled more than a third of the sites, and the pattern I had expected (and secretly hoped to find) was holding up very well. By evening we were more than half-way finished with the sampling, and by this time there was no more drama, just the remaining work needed to complete the sampling. When we departed the Semichi Islands the following evening, I knew what our samples from that area would show.

Amchitka Island is about 24 hours of steaming east of the Semichis. Our travel schedule thus had us arriving at Amchitka in the early evening the following day, too late to begin working. Kiska Island is about eight hours west of Amchitka, so we decided to stop at Kiska for the afternoon and make the transit to Amchitka during the night. Kiska Harbor is where the Japanese scuttled a large number of their remaining naval vessels toward the end of World War II. We were excited by the opportunity to explore sunken wreckage.

There wouldn't be time for sampling at Kiska. But I knew that the island supported an abundant sea otter population, so I wanted to have a look at the surrounding reefs to see whether they were urchin or kelp dominated. I looked at two sites near the mouth of Kiska Harbor and found them to be very much like Amchitka. Samples from Kiska in later years would confirm this impression. But for now we were off to Amchitka, 15 years after my initial work there. I wondered whether it would be the same, and I wondered if whatever we found would prove to be as broadly recurrent as the urchin barrens were in the Semichi Islands. Good weather held, and by the end of the second day we had sampled more than 30 sites at Amchitka. There were no surprises. Everywhere we sampled and everywhere we looked, kelp was abundant and urchins were small and rare.

Next we were off to Adak, about a day's steam east from Amchitka. Adak is in the Andreanof Islands group of the central Aleutian archipelago, almost 500 miles east of where we began the cruise. After the fur trade, Adak had remained otter free until the late 1950s, at which time it was recolonized by the large and rapidly spreading population of otters from the islands immediately to the west. By all accounts, the sea otter population at Adak had been at or near carrying capacity for at least 20 years. I had never before dived at Adak but expected to see reef communities similar to those at Amchitka, even though these two islands are different-looking places above water. Amchitka is comparatively flat and rolling, whereas Adak is extremely mountainous and rugged. But here again there were no surprises. Adak was like Amchitka. Under water, I couldn't have distinguished one island from the other.

We were now 10 days into the two-week cruise and still almost two days of steaming west of Dutch Harbor, the point at which we were to disembark the *Alpha Helix* and fly home. There was little time to dally en route to Dutch Harbor, and little of value to do in the way of fieldwork on our project, given that the status of sea otter populations between Adak and Dutch Harbor was uncertain. Even so, we stopped here and there to dive in places where few or no others had been before.

The vast region between Adak and Prince William Sound was largely unknown to scientific diving. Kathy Ann Miller (a marine botanist from UC Berkeley) was along to collect algae from this poorly known region. She and

I were especially interested in locating the western range limit of bull kelp *(Nereocystis luetkeana),* which abounds along the west coast of North America and, I knew, occurred at least as far west as Dutch Harbor on Unalaska Island. I had often seen drifting plants in the central and western Aleutians, apparently healthy but never attached to the seafloor. We found a large stand of bull kelp on the west end of Samalga Island, westernmost of the Fox Islands and coincidentally the westernmost of the Aleutian Islands to occur on North America's continental shelf. West of Samalga, the Aleutians all are *oceanic islands* rather than *continental islands.* We were quite sure that this marked the western range limit of bull kelp, at least at the time, and published a note on the finding (Miller and Estes, 1989). But I still have no idea why the species' distribution ends so abruptly at that particular place. It was a remarkable adventure through uninhabited volcanic islands, made all the more enjoyable by the knowledge that the goals of our cruise had been achieved.

After returning home from the cruise, I sat down to the task of analyzing the large volume of data from the kelp forest surveys. I had information from more than 150 sites, each of which comprised abundance measurements of the various kelp species, sea urchins, and other common seafloor organisms. For simplicity, I aggregated the kelp counts into measures of total density (numbers of plants per 0.25 square meter) and transformed the urchin counts and size distributions into an estimate of biomass density (grams per 0.25 square meter). I then computed averages of both these measures for each site and plotted the data on a graph, with kelp density on one axis and urchin biomass density on the other (figure 7.2a).

This illustration provides a straightforward view of how the abundances of kelp and sea urchins varied among sites at islands with and without sea otters. Urchin biomass density at the otter-dominated islands (Adak and Amchitka) was consistently low, averaging about 25 grams per 0.25 square meter and ranging from near zero up to 150 grams per 0.25 square meter. Kelp densities at those sites were high but variable, averaging about eight plants per 0.25 square meter and ranging from near zero up to about 17 plants per 0.25 square meter. These same measures differed markedly at the islands that lacked otters. Urchin biomass density at those islands was high but variable, averaging about 325 grams per 0.25 square meter but ranging from near zero to almost 1,000 grams per 0.25 square meter. Kelp densities at these same sites were consistently low for the most part, averaging near zero plants per 0.25 square meter.

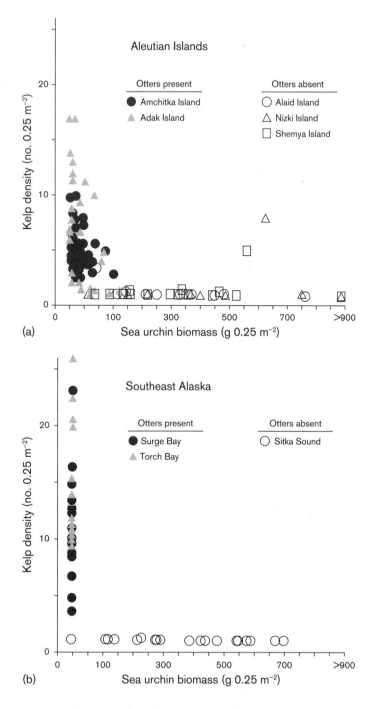

FIGURE 7.2. *State space* plots of kelp versus sea urchin abundance, expressed as number (no.) or grams (g) per quadrat (0.25 m⁻²) for (a) the Aleutian Islands, (b) southeast Alaska, and (c, d) Vancouver Island (a and b are from Estes and Duggins, 1995; c and d are from Watson and Estes, 2011).

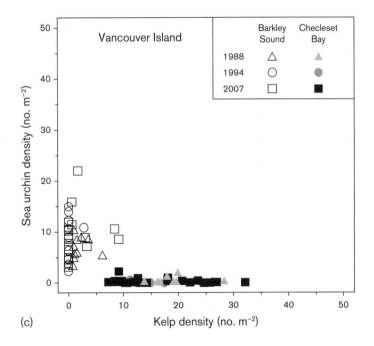

(c)

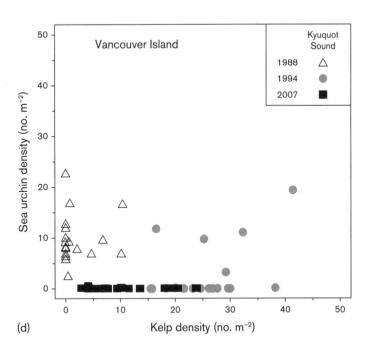

(d)

With the exception of several sites at which both urchins and kelps were rare, these clouds of data points revealed distinct differences between islands with and without sea otters in an objective (unbiased) and statistically robust way. One could easily predict the presence or absence of sea otters from the abundance of kelps and sea urchins at almost any given site. In aggregate, these data demonstrate that the otter–urchin–kelp trophic cascade is a broadly recurrent process across the central and western Aleutian Islands. This is not to say that other factors, such as local oceanographic conditions, do not influence the distribution and abundance of kelp and sea urchins. Indeed, such factors may well be responsible for much of the among-site variation within islands. However, the otter–kelp trophic cascade is apparent through the effects of any and all of these other processes, whatever they might be.

The data obtained in 1987 during our cruise on the *Alpha Helix* demonstrated a generalizable influence of the otter–kelp cascade across the western and central Aleutian archipelago. But as explained earlier, this was only a small portion of the geographic range over which sea otters and kelp forests occurred. What about elsewhere? Although I knew that the trophic cascade occurred elsewhere, an extension of findings from the Aleutians to these other places required a leap of faith that critical scientists would not be willing to make. Independent sampling programs of the sort I had undertaken in the Aleutians were needed to properly assess generality and variation in the otter–kelp trophic cascade elsewhere. As time and opportunity allowed, I would spend much of the rest of my professional life doing just that.

The next opportunity to assess generality and variation in the otter–kelp trophic cascade arose less than a year later, in southeast Alaska. My interest in southeast Alaska stemmed from the reintroduction of sea otters to this region in the late 1960s, which by the late 1980s had created areas with and without sea otters. One such area with otters was the outer coastal region near Cross Sound where David Duggins had documented the otter–urchin cascade a decade earlier. By the late 1980s this subpopulation of otters had spread from Yakobi Island, over Cross Sound to Cape Spencer, northward along the remote outer coast of Glacier Bay National Park, to beyond Torch Bay. We knew that otters had recolonized Torch Bay, because National Park Service biologists had seen them there in large numbers a year earlier. David, who still had close connections with the park service, was interested in knowing how Torch Bay had changed since the time of his dissertation work.

And so was the park service, who provided David with several weeks of ship time on their 65-foot support vessel, the M/V *Nunitak*.

When David invited me to join him, I jumped at the chance. Our plan was to sample Torch Bay with the same protocol we had used in the Aleutians. Time permitting, we would do the same in Surge Bay. This would provide data sets from two areas with sea otters in southeast Alaska. Later that summer I would return to southeast Alaska to sample Sitka Sound, an area that had not yet been recolonized by otters.

The patterns that emerged from the analysis of these data were similar to those I had seen in the Aleutians, if not even more extreme (figure 7.2b). Everywhere we looked, the shallow reefs of Torch and Surge bays were festooned with kelps. Although we saw the occasional small sea urchin, not a single urchin was found in our more than 600 randomly placed plots in these two locations. As in the Aleutians, kelp densities were high but variable. Although David had seen the impact of sea otters from our cruise across the Aleutians and his own earlier graduate work in southeast Alaska, he was nonetheless astonished by how much Torch Bay had changed since he had worked there a decade before.

Sitka Sound in 1987 was the mirror opposite. Everywhere I looked, the shallow reefs were covered by red and purple sea urchins. Although I saw the occasional kelp here and there as I was looking about or traveling to the sample sites, not a single kelp plant occurred in any of 600 randomly placed plots that were spread across the 30 randomly selected sample sites in Sitka Sound. As in the Aleutians, urchin biomass densities were high but variable.

I returned to southeast Alaska on several occasions over the next two decades to resample the same sites at these three areas. The Torch and Surge bay sites have remained largely unchanged—adorned with kelps and lacking sea urchins almost entirely. Sea otters spread into Sitka Sound during the early to mid-2000s and, not surprisingly, the sample sites had changed markedly when I resurveyed them in 2009. At that time, the abundance and distribution of kelps and sea urchins on Sitka Sound reefs were indistinguishable from those in Torch and Surge bays.

A third opportunity to assess generality and variation in the otter–kelp trophic cascade existed on the west coast of Vancouver Island. As in southeast Alaska, sea otters were reintroduced to Vancouver Island following their extirpation in the maritime fur trade. The reintroduction site was Checleset Bay, on the island's northwest coast. A colony was well established in this area by the late 1980s and had begun to spread north and south.

Jane Watson, my former graduate student, used the recolonization and spread of otters along the west coast of Vancouver Island to measure the generality and variation of their influences on kelps and sea urchins, much as Dave Duggins and I had done in the Aleutians and southeast Alaska. Jane's work, begun in earnest in 1990, was done at three locations: Checleset Bay, where otters were well established and were expected to remain so; Kyuquot Sound, where otters were absent but were expected to become reestablished in the near future; and Barkley Sound, where otters were also absent but were expected to remain so for quite a few more years.

Except for a few species differences that inevitably occur between one geographic region and another, Jane's data from Checleset Bay were similar to those we had gathered from the otter-dominated sites in southeast Alaska (figure 7.2c). Kelp densities were high but variable, and urchins were largely absent from all her sample sites and sample plots. This pattern remained unchanged in Checleset Bay over the 20-year period of her study. Her data from Barkley Sound (figure 7.2c) were similar to those we had obtained from Sitka Sound in the late 1980s, before the area was recolonized by sea otters. Here again, the overall pattern remained largely unchanged over the 20-year period of Jane's study. Her sites in Kyuquot Sound, by contrast, changed markedly during this same period following the otter invasion (figure 7.2d).

One of Jane's more remarkable accomplishments is that she was in Kyuquot Sound to observe the reef phase shift as it happened in real time. She watched the seafloor beneath the first invading otters as they fed on the abundant large red urchins, and she saw the remaining living urchins flee from the discarded remains of their dead fellows after they were dropped to the seafloor by foraging otters, thus creating urchin-free patches that were quickly colonized by kelps. For a short while, the reef system was a patchwork mosaic of kelp and super-dense urchins. But it wasn't long before the otters found and consumed all the urchins, at which point the system was transformed into an unbroken kelp forest.

Collectively, these three data sets from the west-central Aleutians, southeast Alaska, and British Columbia establish a broad generality to the otter–kelp trophic cascade from the western Aleutians to the Strait of Juan de Fuca. Virtually everywhere we looked, the presence or absence of sea otters was a reliable predictor of whether the seafloor would be covered with kelps or deforested by urchin grazing. Our samples, of course, comprised only a tiny fraction of the thousands of square miles of shallow seafloor habitat across the Pacific Rim. But given the wide range of locations over which the samples were obtained, the nature of the sampling protocol, and the consistency of the findings, the results give us every reason to believe that the concept of the otter–kelp trophic cascade applies to most of the shallow reef habitat across this vast region. Sea otters are clearly more than "just another brick in the wall."

EIGHT

A Serpentine Food Web

ONCE I UNDERSTOOD THE OTTER–KELP trophic cascade, it was easy
to imagine knock-on effects to other species and ecological processes. The
underlying logic was simple. If otters affected kelp abundance, and if kelp
forests provided food and habitat for other species, then these other species
must be indirectly influenced by sea otters. The only possible flaw I could see
in that logic arose from uncertainty over the relative importance of kelp production. Kelp (macroalgae) and phytoplankton (unicellular algae) are the
two main groups of primary producers in higher-latitude coastal marine ecosystems. If kelp production were low compared with that of phytoplankton,
or if phytoplankton production had increased to compensate for the loss of
kelp production in the absence of sea otters, then the otter's broader ecological influence through the coastal food web might be diminished. But I knew,
from earlier work by others, that kelp growth and production were remarkably high, so I was almost certain from the outset that the sea otter's influence
on coastal ecosystems would prove to be strong and far reaching. The challenge was in identifying and demonstrating those influences.

In order to more clearly envision the dynamics of this complex web, I
needed to do three things. First and foremost, I needed to envision a more
comprehensive food-web topology. In other words, I needed to understand
the nature of direct species interactions and how these interactions were
linked in a food web. Although all ecosystems are ultimately fueled by *primary production,* it doesn't necessarily follow that primary production drives
the abundance of species. The topology of food webs is more complex than
this, owing in large measure to the complexity of species interactions (e.g.,
the abundance of a species is not always dictated solely by food but often also
by its predators, competitors, mutualists, and the limiting influences of the

physical environment), the spatial and temporal scales over which these inter-actions operate, and the ways in which direct interactions between species are linked to define pathways through food webs.

Second, I needed a more focused view of what to look for. That is, I would need to formulate explicit hypotheses about how sea otters might influence other species and ecological processes that, in turn, linked back to the otter–kelp trophic cascade. If I got the topology wrong, then the hypotheses were meaningless and I would be doomed to failure from the outset, no matter how powerful and robust the ecological influences of sea otters might really be. And third, I needed a way of testing these hypotheses in a simple but compelling way.

Searching for knock-on effects of the otter–kelp trophic cascade was something I had wanted to do from early in my research career. The main difficulty was finding time to pursue this interest, already having all that I could manage (and more) with the otters, urchins, and kelps. The more far-reaching looks into food-web dynamics came to fruition largely through the efforts of my students and colleagues. In some cases I passed the seed of an idea on to one of these people to develop and run with. In other cases that spark of insight came from someone else's mind.

SEA OTTERS AND SEAGULLS

Our first glimpse into the knock-on effects of the otter–urchin–kelp trophic cascade was provided by glaucous-winged gulls *(Larus glaucescens)*. These seagulls are common and widespread across the Aleutian archipelago, living and foraging primarily in coastal marine habitats. The inspiration to explore the potential interplay between glaucous-winged gulls and the otter–kelp trophic cascade came from David Irons shortly after we began working together on Attu Island in 1976. Dave, who knew the otter–urchin–kelp story, had been watching gulls forage in the intertidal zone during low tides and noticed that they were consuming large numbers of intertidal inverte-brates, especially urchins, mussels, chitons, and limpets. He further noticed that the gulls quickly swallowed their smaller to mid-sized prey intact but were incapable of doing this with the largest urchins. Instead, the gulls gained access to the nutritious entrails of these large, energy-rich urchins either by pecking the test away around its oral aperture or by breaking the test apart by air-dropping the urchin on the rocky intertidal bench. The difficulty for gulls in pecking out the oral aperture is that it takes a lot of time; the difficulty with air-dropping is that it attracts other gulls, which then steal the urchin's edible viscera when it splatters about on the rocks.

Foraging gulls were always entertaining to watch. But Dave was beginning to wonder about the economics of gull foraging: what species and sizes of invertebrates they selected; whether those choices were governed by maximization of energy gained; and how these factors were influenced by sea otters. Dave's initial observations came from Massacre Bay (where otters were absent at the time), and he suspected that much of what he was seeing in the gulls' foraging behavior was a product of the absence of sea otters. A more rigorous study of prey selection and foraging economics in glaucous-winged gulls was a natural, and Dave did just that, working with our mutual friend and colleague Bob Anthony as a graduate student at Oregon State University. I would support the project and serve on Dave's thesis committee, and Bob and I would help Dave with the fieldwork on Attu.

Our main premise was simple and straightforward. We imagined that depression of invertebrate populations by foraging otters would force co-occurring gulls to consume a more diverse, or even entirely different, array of prey species. To test this hypothesis, we chose to compare the diet and foraging behavior of gulls among three areas: Massacre Bay, where otters were absent at the time; Chichagof Harbor, where otters occurred but were well below carrying capacity; and Amchitka Island, where otters had been at or near carrying capacity for decades. We characterized the overall diets of the gulls by watching focal animals forage at low tide and by sorting through regurgitated pellets for prey remains at gull loafing sites.

The gulls' dietary differences among these three sites were striking, more or less as we hypothesized they would be, but with a few interesting twists we had not anticipated. On Attu the gulls fed mostly during low tides, when intertidal invertebrates were exposed and available to them. In Massacre Bay, they consumed a narrow range of moderate to large-sized invertebrates, mostly urchins and chitons during spring low tides but including more mussels and barnacles during low *neap tides*. We documented the same general pattern at Chichagof Harbor, except here the gulls consumed a wider array of invertebrate species. The gulls' diets at Amchitka differed radically from the patterns we had seen on Attu. At Amchitka the gulls fed almost entirely on fish.

In this simple study of gull foraging behavior, we employed the same fundamental method I had used earlier to describe the otter–urchin–kelp trophic cascade. That is, we compared otherwise similar locations with and without sea otters to explore the ways in which gulls might be influenced by the effects of otters on the coastal ecosystem. The results were basically what, in retrospect, we should have expected to see (figure 8.1). Where otters were absent, gulls fed on the abundant sea urchins and other intertidal invertebrates that developed in their absence. As otters recolonized coastal habitats and began reducing the abundance and size of these intertidal invertebrates, the gulls responded by diversifying their diets. Where otters had recovered to their historical carrying capacity, reduced invertebrates to very low levels, and effected a recovery of kelps through the otter–urchin–kelp trophic cascade, the gulls forsook invertebrates almost entirely for a diet of fish (Irons et al., 1986).

SEA OTTERS AND SECONDARY PRODUCTION

Any thoughtful conceptualization of how otters might influence other species and ecological processes via the otter–kelp trophic cascade necessarily leads to the question of how kelp forests themselves influence other species and ecological processes. My own thinking on this front was fuzzy until I teamed up with David Duggins and Si Simenstad in the mid-1980s (figure 8.2). David and Si had the question well in mind and envisioned such influences occurring through one or some combination of three specific processes: changes in production, the creation of biogenic habitat, and altered flow. Their thinking was clear and simple: Production effects would occur through photosynthesis—the transformation of CO_2 by kelps into glucose and other derived forms of organic carbon that could be utilized by primary consumers (such as sea urchins and bacteria) to fuel the ecosystem. Habitat effects would occur through creation of the three-dimensional structure of the kelp forest above the seafloor, which became a place for other species to live. Much as forest birds disappear following a clear-cut because of habitat loss, their resulting inability to take refuge from predators, and the compromise of myriad other life functions, kelp forest animals might be expected to disappear after these forests are mowed down by foraging urchins. Kelp adds surface area to the water column, and surface area creates friction—hence, the third mechanism David and Si envisioned for kelp's influence on the ecosystem was the altered patterns of water flow from wave surge and coastal currents. The typical calm of a terrestrial forest floor during otherwise windy conditions results from the same physical principles.

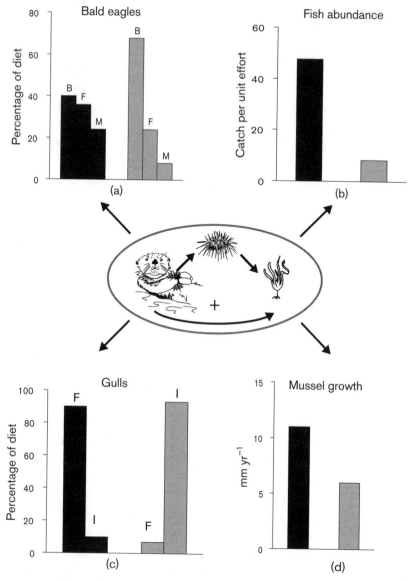

FIGURE 8.1. Four indirect effects of the sea otter–urchin–kelp trophic cascade (depicted in the figure's center): (a) bald eagle diet (B = seabirds; F = fishes; M = marine mammals), (b) kelp forest fish density, (c) glaucous-winged gull diet (F = fishes; I = invertebrates), and (d) blue mussel growth rate (figure modified from Estes et al., 2011). Data are from areas where otters were present (black bars) or absent (gray bars).

FIGURE 8.2. From left to right: David Duggins, the author, Jane Watson, Charles "Si" Simenstad, and Patrick Hassett, outside the LORAN A station near Casco Cove on Attu Island in 1986. (Photo by Charles A. Simenstad)

Here again, the specific hypotheses were simple and easy to test by contrasting islands in the western and central Aleutians with and without sea otters. We would approach the issue on two fronts: by using stable carbon *isotopes* as a signature of kelp-derived organic carbon in other species, and by measuring growth rates of suspension-feeding invertebrates. Kelp has a stable carbon isotopic signature that is distinct from that of phytoplankton, and we thus hypothesized that a comparison of the same suite of species between islands with and without sea otters would reflect this difference in a consistent manner. From our general understanding of kelp forest natural history, we knew that the rapid growth rates of kelp were not manifested so much in increasing plant size as in the sloughing of tissues into the water column in the form of particulate organic carbon and dissolved organic carbon. If greater kelp abundance at islands with sea otters increased the coastal ecosystem's overall net primary production, we expected to see this effect in the growth rates of animals that directly consumed particulate organic carbon—the suspension feeders.

We chose to test this hypothesis in the following manner. First we obtained recently settled mussels and barnacles from an area near the University of Washington's Friday Harbor Labs on San Juan Island in Puget Sound. We selected two common species that ranged widely from the west coast of Washington through the Aleutian archipelago. We then planted out a few individuals from this common source pool of mussels and barnacles to various islands with and without sea otters in the western and central Aleutians. The mussels and barnacles would need time to grow, so we planted them out in the summer of 1986 and collected them a year later.

This was easy to do in concept, but we worried that the mussels and barnacles might be eaten by predators or washed away during the violent storms that were bound to occur in the interim between when the animals were planted out and when they were collected a year later. To protect them from predators (or just about anything else that could harm them), we placed the mussels and barnacles in stainless steel cages that were specifically designed and fabricated for our study. To prevent the cages from being washed away, we affixed them to the seafloor by drilling into the rocky substrate, securing threaded stainless steel studs in the drilled holes, and then bolting the cages firmly into the underlying bedrock over the threaded studs. We planted out six sets of barnacles and mussels in the intertidal zone and six additional sets at depths of 20 to 30 feet in the adjacent subtidal zone at each of four islands. Two of these islands (Adak and Amchitka) supported abundant sea otter populations; at the other two, otters were rare (Attu) or absent (Shemya).

We hoped the cages would survive the year, but we worried they might not and expected that some would be lost. But in fact we were able to relocate and recover all 48 cages. Moreover, the results were just as we had expected: both the barnacles and the mussels grew two to three times faster at Adak and Amchitka than at Shemya and Attu (figure 8.1; for simplicity, only the mussel data are shown). We took this as strong evidence for a significant increase in net primary production from the otter–kelp trophic cascade (Duggins et al., 1989).

SEA OTTERS AND FISH

Much like birds in terrestrial forests, fishes are among the more conspicuous denizens of kelp forest ecosystems. From early on in my studies of sea otters and kelp forests, I suspected there were significant ecological linkages between the otter–urchin–kelp trophic cascade and kelp forest fish species. This was only logical if kelp was in any way important to the fish.

There were hints of such linkages in my earlier studies. One was from the sea otter's diet, which varied markedly between Amchitka Island, where otters and kelp were abundant, and Attu Island, where at the time otters had just reestablished themselves, their numbers were still low, and kelps were sparse. The otters of Attu fed almost exclusively on sea urchins and other invertebrates, whereas more than half of the diet of the otters at Amchitka consisted of fish (Estes et al., 1982). Rock greenling, a common kelp forest fish in the Aleutian Islands, was especially common in the otters' fish-dominated diet at Amchitka. I knew this because male greenlings are large and easily identified by their distinctive red and black skin and the even more distinctive bright blue of their flesh. David Irons's work on glaucous-winged gulls also hinted at a similar process. The gulls of Attu fed largely on invertebrates, whereas those at Amchitka fed mostly on fish.

Despite this simple logic and the indirect supporting evidence for a link between fish species and the otter–kelp trophic cascade, I lacked the necessary field data to demonstrate a relationship between otters, kelp forests, and kelp forest fishes in a manner that would be sufficiently convincing to the scientific community. Part of the difficulty was methodological. In contrast to sea urchins and kelps, which are easy to sample because of their abundance and lack of mobility, fishes are less abundant and much more mobile. Another difficulty in measuring fish populations in the same general way I had measured urchins and kelp is that fish are often either attracted to or repelled by divers. Rock greenling are especially difficult to count because the females are extremely shy whereas the males, which guard their nest sites, are aggressive toward divers. In my thousands of hours spent diving in the Aleutians over the decades, nearly all the female greenling I saw were little more than a blur as they raced away. By contrast, most of the adult males either hovered over the egg masses they were guarding or attacked when I approached their nests. Furthermore, fish were more difficult to see amid the dark and constantly moving background of a kelp forest than over the lighter, stationary background of an urchin barren. Although I conducted fish counts along transects through kelp forests and over urchin barrens at various times and places, and the fish seemed to be more abundant in kelp forests than in the urchin barrens, the data were too messy to conclude very much about population densities.

As a graduate student at Amchitka Island in the late 1960s and early '70s, Si Simenstad sampled fish by setting a trammel net on the reef for several hours. Catch per unit of effort—in this case the number of fish caught per length of net per time—provided an index of relative abundance. Si had continued to use this method to sample reef fish populations at various times and places across the Aleutians from the mid-1970s to the mid-1980s. His collective data set included information gathered from multiple sites and in multiple years at Adak, Amchitka, Shemya, and Attu during a period in which otters were abundant on Adak and Amchitka and rare or absent on Attu and Shemya. The catch-per-effort measures for rock greenling were notably higher at Adak and Amchitka than at Shemya and Attu, from which we inferred that the otter–kelp trophic cascade likely enhanced rock greenling populations. Si never published these data, in part over concern for his inability to distinguish between effects of the otter–kelp cascade on one hand and potential confounding effects of other possible differences among the islands on the other. These data lay fallow in the archives of Si's lab at the University of Washington until the early 2000s, when two events conspired to resurrect the question and move forward our efforts to resolve it.

One of these events was the arrival of the right graduate student in my lab. I had gotten to know Shauna Reisewitz in 1999 as one of my son Colin's high school science teachers, and I liked her from the moment we met. So when she approached me several years later about doing a master's degree at the University of California (UC) Santa Cruz, I was warm to the idea and introduced her to the sea otter–kelp forest fish question in hopes that she might take it on as a thesis project. Shauna liked the idea, and I had a funded contract for work in the Aleutians that would provide the necessary ship and field support to revisit and resample fish at Adak, Amchitka, Shemya, and Attu. It was decided quickly: Shauna would join my lab and take on the project.

The second event, which made the fish question tractable and more intriguing, was the collapse of sea otter populations across the Aleutian archipelago. By the late 1990s, kelp forests on the reefs surrounding Amchitka and Adak had followed suit by transitioning to extensive sea urchin barrens (discussed further in chapter 10). I had documented the phase shift at these two islands. From similar sampling efforts at Shemya and Attu, I also knew that these systems had remained unchanged as sea urchin barrens over the same period. The data in support of these conclusions were extensive and unequivocal. Inasmuch as Si had fish catch-per-effort data from all four of these islands from before the sea otter's collapse, the resampling of fish populations at each of the four islands would provide a simple and obvious test of the otter–kelp–fish hypothesis. If the catch-per-effort measures at Shemya and Attu remained unchanged whereas those from Adak and Amchitka declined to these same values, we would be able to infer with reasonable confidence that the hypothesized effect of the otter–kelp trophic cascade on kelp forest fish was indeed correct and real. Otherwise, if the earlier differences between Shemya–Attu and Adak–Amchitka remained unchanged, we would infer that the difference in fish abundance between these two sets of islands was caused by something else.

Shauna joined me on a research cruise across the central and western Aleutians in the summer of 2000, during which she systematically repeated Si's sampling protocol at Adak, Amchitka, Shemya, and Attu. She worked like a Trojan and was able to complete the fieldwork as hoped and planned. The data confirmed our hypothesis of a positive association among otters, kelp, and rock greenling (figure 8.1). Before the sea otter–kelp forest collapse, overall catch per unit effort for rock greenling was about eightfold higher at Amchitka and Adak than at Shemya and Attu. After the collapse, greenling capture rates remained largely unchanged at Shemya and Attu, whereas those at Amchita and Adak declined to levels we had seen throughout at Shemya and Attu (Reisewitz et al., 2006). A few years later Russell Markel, a graduate student at the University of British Columbia, found a similar strong positive association among sea otters, kelp, and rockfish along the southwest coast of Vancouver Island (Markel and Shurin, in press).

I had long imagined a connection between sea otters and kelp forest fishes through the otter–kelp trophic cascade. But science is based on data, not supposition. The collective findings of Shauna's and Russ's independent research resolved any real question in my mind between supposition and fact by providing clear and unequivocal evidence for an important ecological linkage between sea otters and kelp forest fishes.

SEA OTTERS AND SEA STARS

I had long imagined that the sea otter's principal influence on other species and ecological processes would be through the otter–kelp trophic cascade. But sea otters consume numerous other species besides sea urchins, and the resulting array of direct predator–prey interactions creates alternative pathways by which their ecological effects might spread through the coastal ecosystem. One such possibility that came to mind near the beginning of my field studies on Attu involved the interaction between sea otters and sea stars.

No fewer than a dozen species of sea stars live on shallow reefs in the Aleutian archipelago. Many of them are predators, feeding on other species of sessile or weakly motile invertebrates such as mussels, barnacles, snails, urchins, and other sea stars. Several of the sea star species were large (a half meter from ray tip to ray tip) and abundant on the seafloor around Attu, especially in Massacre Bay, where otters were absent. I also knew from Bob Paine's then recent work on *Pisaster ochraceus* (the ochre star) in Washington that sea stars could be important predators, limiting populations of their preferred prey (intertidal mussels in Paine's studies) and thereby influencing the structure and diversity of the entire rocky intertidal community. I knew from my earlier field studies at Amchitka that otters ate the occasional sea star, so I wondered whether sea otter predation on sea stars might not influence the ecosystem by reducing sea star populations and thereby altering whatever predatory effects these sea stars were having on the system. The anticipated recovery of otters on Attu and their spread into Massacre Bay would provide a means of answering that question.

The overall approach of documenting change with the spread of otters into Massacre Bay was simple enough. But I needed to make the questions more explicit, and to do this I needed to know more about sea stars. In particular, I needed to document their abundance and population structure in this otter-free habitat around Attu. I also needed to know what the sea stars around Attu were eating and the extent to which these predator–prey interactions were controlling or otherwise influencing their prey.

I began delving into these issues in 1979 by running 25-meter-long transect tapes across the seafloor and measuring the species, size, and prey of every sea star I found within a meter-wide band along the transect. Working with another diver, I found that I could complete one transect on a single tank of air. As in the sampling of sea urchins and kelp, I began each transect wherever the anchor of my dive skiff happened to fall, swimming in one direction and then the other along the depth contour of the anchor's start point. My time in the field was limited, and there was other work to be done. So I arbitrarily decided to sample five transects at each of three locations on the western side of Massacre Bay. I thought these locations were more or less representative of the surrounding region in regard to the distribution and abundance of sea stars. But more importantly, any effects of sea otters ought to become apparent with changes that occurred as otters spread into Massacre Bay.

The greater challenge was in determining the ecological effects of sea star predation and how this might change with the addition of otters to the system. Bob Paine had explored the effects of sea star predation through an experiment in which he removed all sea stars from one area in the rocky intertidal zone, left a nearby area unchanged, and then watched the associated communities diverge through time. I couldn't do that on Attu because the sea stars moved slowly but extensively across the seafloor and I would be there only a month or two each summer to maintain the experiment. Therefore, I used a different approach, better suited to the limitations of my own field situation, that would provide interesting data on the interaction strength between sea stars and their invertebrate prey. I decided to focus on just two common sea star prey: mussels and barnacles. Although I rarely encountered either of these invertebrates on the shallow subtidal reefs, I knew they could live in the subtidal community and that they were common in the adjacent rocky intertidal zone. My plan was to transfer rocks with mussels or barnacles already living on them from the intertidal zone onto the adjacent subtidal reef. I would assess the impact of sea star predation by placing some of these rocks in cages to which sea stars were prevented access, others on the open seafloor with free access by sea stars, and then measuring differences in survival between caged and uncaged mussels and barnacles over the course of a summer's field season.

While it was simple in concept, there were several shortcomings and potential pitfalls to my experimental design. For one, I couldn't be sure that moving the mussels and barnacles from the rocky intertidal into a new and different subtidal environment wouldn't influence them in some way beyond the exposure to sea star predation. If these animals were compromised by either their new environment or the stress of being moved, those compromises might also make the uncaged individuals more vulnerable to predators. In an effort to assess any such effects, I added a third treatment to the experimental design. I returned some of the mussel- and barnacle-covered rocks to the intertidal sites from which they had been collected and monitored their survival.

As it turned out, the mussels and barnacles used in these intertidal controls and those moved to the subtidal cages survived at similarly high rates. I concluded from this that life in the subtidal environment, in itself, did not compromise the mussels' or barnacles' welfare and, in the interest of time, discontinued the intertidal controls. A second concern was that any difference in survival between the caged and uncaged mussels and barnacles might be caused by something other than sea star predation—a fish predator, for example. Sea stars feed by everting their stomachs over their prey and digesting them, leaving what remains of the prey's exoskeleton in a state that is diagnostically different from what crushing and drilling predators (such as fish, snails, and octopus) leave of their prey. Once the experiments were under way, I visited them daily and could also see firsthand that all (or at least most) of the predation was by sea stars.

This first phase of the study was finished in 1983, at which time sea otters were poised to spread into Massacre Bay. They did so in 1986, and by the early 1990s large numbers of otters had been living and feeding in the area for several years. I had been watching the system in the interim, and the effects of otter predation on sea stars were obvious. Once otters arrived in Massacre Bay, I began to notice some of the larger stars with missing arms, something I hadn't seen before. Soon thereafter, the larger individual stars disappeared entirely. It was time to resurrect the field research by remeasuring the sea star populations and repeating the mussel and barnacle translocation experiments.

I didn't have time to do this work myself, so again I recruited a student to help out. Ken Vicknair, a recent undergraduate from UC Santa Cruz and a strong diver who had worked with me in California, was ideally suited for the task. He was interested in sea stars and knew their systematics and natural history very well, largely from the undergraduate mentoring he had received from John Pearse, a UC Santa Cruz professor and expert on marine invertebrates. Most importantly, Ken was keen to do a master's degree and his excitement over the possibility of fieldwork in the Aleutians was obvious. So it was quickly decided. I would send Ken and two field assistants to Attu for the summer, they would repeat the sea star measurements and experiments, and Ken would use these data and those I had gathered earlier for a master's thesis. There were a few minor bumps along the way, but in the main this worked out as well as I could have hoped.

Our data differed strikingly between the periods before and after sea otters recolonized Massacre Bay. The maximum diameter of the largest sea stars declined by about 50 percent, and the overall biomass density of sea stars declined more than tenfold (Vicknair and Estes, 2012). The survivorship patterns of mussels and barnacles in the caged controls remained unchanged from 1983, but survival increased markedly for uncaged animals on the open seafloor, indicating that the interaction strength between sea stars and their invertebrate prey had declined.

The results of these studies on the ecological interplay between sea otters and sea stars led to three conclusions. First, the limiting influence of sea otters on sea star populations was strong and roughly comparable in magnitude to the limiting influence of sea otters on sea urchins. Second, the reduction in sea star populations by foraging otters resulted in a reduction in the interaction strength between sea stars and their prey. Finally, given the well-known direct and indirect effects of sea stars on other species and ecological processes, the sea otter–sea star interaction probably has ecological influences that extend well beyond those that Ken and I saw and measured.

SEA OTTERS AND BALD EAGLES

Bald eagles are common in the Aleutian Islands, where the average density of nesting pairs is about one per 3 or 4 kilometers of shoreline. I had long suspected an ecological link between sea otters and eagles because the latter feed largely on marine resources, which otters profoundly influence through the otter–urchin–kelp trophic cascade. But for years, all I could do was imagine what these effects might be—because, while I could generate logical hypotheses, I had no obvious means of putting those hypotheses to experimental tests. I couldn't do with eagles what Dave Irons had done with gulls, because the bald eagle's geographic range mysteriously ended at Kiska Island. I've wracked my brain in an effort to understand why the eagle's range ends so abruptly there, but I've never figured it out. Perhaps the Near Islands are somehow unsuitable habitat for eagles, or perhaps eagles never recolonized the Near Islands after the Pleistocene. Or it may be that the lack of eagles in these islands is an ecological casualty of the absence of otters. Whatever the explanation, the absence of eagles in the Near Islands prevented a comparison of the natural history and ecology of these birds between areas with and without sea otters.

The opportunity to learn more about the ecological interplay between sea otters and bald eagles began to develop in the early 1990s with a National Science Foundation–funded study of sea otters at Amchitka Island, which included several weeks of ship support aboard the *Alpha Helix* each year. During the project's first year, I used the *Alpha Helix* as a staging platform for the work at Amchitka (see chapter 10). In the following two years I used the ship to obtain information on the otters and coastal ecosystems of other islands. The cruise plan during these later years was simple. We would begin at Adak and sail west to Attu, then turn east and sail back to disembark at Dutch Harbor, steaming at night and stopping during the days to visit some

FIGURE 8.3. Bob Anthony, measuring prey remains from a bald eagle nest in the Aleutian Islands. (Photo by Norman S. Smith)

of the larger islands along the reach of the Aleutian archipelago. There were several unoccupied berths, which I wanted to make available to other scientists who might both benefit from visiting the Aleutian Islands and enrich our program by what they would learn. Two long-standing questions loomed in the recesses of my mind. Why did the bald eagle's range end so abruptly at Kiska Island, and how might eagles be influenced by the otter–urchin–kelp trophic cascade? I invited Bob Anthony (figure 8.3) to join us in trying to answer these questions.

Bob was a lifelong friend and respected colleague who had studied bald eagles in the Pacific Northwest and interior Alaska for years. He was especially interested in how the life history and ecology of eagles varied among the diverse ecosystem types in which the species occurred. Bob had studied eagles in forest and riverine systems but wanted to expand this body of work to include maritime populations. The Aleutians intrigued him because of both the abundance of eagles and the ease with which their nests could be accessed and observed in this treeless landscape.

In the beginning, Bob's eagle work was largely descriptive, focusing on reproductive success, diet, and contaminants in shells and unhatched eggs from the nests he visited. From his perspective, these data were valuable as points of contrast with the other systems he and others had studied elsewhere. The treeless landscape was a particular boon to his efforts because it allowed him to walk into and observe as many nests in a day as he might be able to visit in weeks of fieldwork in places where access required the highly technical and often dangerous climbing of tall trees.

Bob's early data were obtained during the summers of 1993 and 1994, at which time sea otters were in the early stage of decline but still abundant and, as revealed by our diving surveys of the coastal reefs, still living in kelp-dominated ecosystems. This data set was of interest to Bob but, by itself, did little to answer the question of how eagles were linked to the otter–urchin–kelp trophic cascade.

The opportunity to answer that question arose a decade later, by which time the otter population had collapsed across the Aleutian archipelago and its reef ecosystems had shifted from kelp forests to sea urchin barrens. All that remained in our documentation of indirect effects of sea otters on eagles was to resurvey the eagles and their nest sites. If the patterns remained unchanged from those of the early 1990s, we would conclude that otters in fact had very little effect on eagles. But if the patterns had changed, and especially if they changed in ways that were consistent with what we already knew about the otter–urchin–kelp trophic cascade, then we could be reasonably certain of an ecological link between otters and eagles.

During 1993–94, Bob and his field assistants surveyed 179 eagle nests on four islands: Adak, Amchitka, Kiska, and Tanaga. During the summers of 2000–02, he resurveyed most of the same nests and 155 others. The overall difference between these two periods was striking, and consistent with what we might have expected given our knowledge of the otter–kelp trophic cascade (Anthony et al., 2008). During the early period, when otters still abounded at these four islands, the eagles fed on a diverse array of marine mammals (including sea otter pups), fish (including rock greenling and other kelp forest fishes), and an array of seabirds. These three prey groups made up roughly equal proportions of the eagles' overall diet. In the latter period, after otters had become scarce and the coastal ecosystem had shifted to the urchin-dominated phase state, the eagle's diet narrowed and became more dominated by seabirds (figure 8.1). The reduction of marine mammals in the eagles' diet was largely a consequence of the disappearance of sea otters, and the reduction of fish was consistent with the kelp forest's collapse and associated declines of kelp forest fishes.

With the discovery of so many indirect effects of the otter–urchin–kelp cascade, I thought I had a reasonably broad view of this phenomenon and a good general sense of how and where to look for additional indirect effects. My overview was largely restricted to the interactions among species, however, and devoid of thinking about how the cascade might interact with the physical environment. In the late 1970s George Jackson (a physical oceanographer from Scripps) had shown a strong attenuating influence of kelp forests on nearshore waves and currents. Given the otter's influence on kelp abundance, there was little doubt that otters acted further to reduce coastal current velocity and wave force. However, that process defined the horizon of my view of how sea otters might influence their physical environment. Looking back on it now, I consider my somewhat myopic view then a result of the same set of blinders that so many ecologists have inherited from the traditions of ecological thinking—namely, the idea that interactions between the physical and biological elements of the environment are mainly a one-way street.

My thinking on this front began to change after meeting Chris Wilmers in 2008. Chris, a newly hired assistant professor at UC Santa Cruz, was doing what any smart junior faculty member should do, which was talking with his new colleagues about their research programs and possibilities for collaborations. As a doctoral student at UC Berkeley and a postdoc at UC Davis, Chris had worked on carnivore ecology and climate change, and he was especially interested in the interplay between the two. Inevitably, these interests had led him to think more extensively about atmospheric carbon dioxide and the *carbon cycle* than I ever had. To my mind, the well-known increase in atmospheric carbon dioxide and the associated influences on global heat retention and ocean acidification were mostly of interest as drivers of future environmental change. Chris saw it differently and asked if I had ever considered how the otter–urchin–kelp trophic cascade might influence atmospheric carbon dioxide. I hadn't, but my immediate reaction was that any such effects were surely minuscule.

Looking back, I think that my skeptical reaction to Chris's question came from two dimensions of my thinking at the time, one of which was rational but the other much less so. The irrational element was my mind-set that biotic interactions had relatively little influence on the physical environment.

Although many scientists had known for a long time that this wasn't the case, I had my own interests and agendas and just hadn't paid enough attention to what these people had learned or its implications for my understanding of nature. On the rational side, I simply couldn't imagine that kelp forests, which occurred over such a small portion of Earth's surface, could influence an atmosphere that is global in extent. Chris agreed on this latter point but didn't think that a global scale was the only one of interest. From his perspective, the more important question was smaller in scale. That is, if we considered just that portion of the atmosphere over the North Pacific Ocean's coastal zone, would the otter–kelp trophic cascade matter? Like all photosynthesizing plants, kelp takes up carbon dioxide from the surrounding environment. The initial question was thus a simple one. What was the difference in carbon dioxide uptake between a world with and one without sea otters, and what proportion of the carbon dioxide in the overlying atmosphere did that difference comprise?

An answer to that question required the following information. First, we needed to know the area of habitat within which sea otters and kelp forests could occur across the North Pacific Ocean and southern Bering Sea. This was determined by taking the total area between the low tide line and the 20-*fathom* depth contour (the approximate depth limit of kelp growth under ideal conditions) and then using various available databases to estimate the proportion of that area underlain by rocky reefs (the actual habitat type on which kelps can grow). Next we needed to know the difference in kelp biomass density between rocky reef systems with and without sea otters. We used the reef survey data from randomly selected locations with and without sea otters in the Aleutian Islands, southeast Alaska, and British Columbia to make that critical calculation (see chapter 7). As it turned out, the values were quite similar in all three regions, indicating that they could be applied as a single proxy for the influence of sea otters on kelp biomass across the entire span of coastal habitat from the western Aleutians (including Russia's Commander Islands) to the southern end of Vancouver Island. We chose to limit the analysis to that region because we didn't have comparable data on kelp abundance between habitats with and without sea otters from elsewhere. Next, we needed to know the carbon content of kelps, which was easily measured in the laboratory. Finally, we needed to know the carbon content of the overlying atmosphere. Since atmospheric carbon is well mixed and broadly uniform across the globe, this value was easily derived from published data on atmospheric carbon concentration and a calculation of the atmospheric volume overlying the sea otter's coastal habitat.

The results were eye-opening. We estimated that the otter–urchin–kelp trophic cascade would result in storage of 4.4–8.7 teragrams of carbon in just the living kelps. That's a lot of carbon. For perspective, 4.4–8.7 teragrams represents 5.6–11.0 percent of the total carbon in the overlying atmosphere, or 21–42 percent of the increase in carbon in this same volume of atmosphere that has occurred since the beginning of the industrial revolution. This amount of otter-generated organic carbon would have a potential net worth in futures of $205–408 million (U.S. currency) on the *European Carbon Market* (Wilmers et al., 2012).

The potential effect of sea otters on atmospheric carbon sequestration and storage becomes even more interesting when considered from the perspective of kelp production and carbon flux. This is because the annual net primary production of kelp forests is roughly three to five times their standing biomass. In other words, the difference in annual absorption of atmospheric carbon from kelp photosynthesis between a world with and a world without sea otters is somewhere between 13 and 43 teragrams. These values are potentially meaningful on a global scale, depending on the fate of that carbon after it exits the living kelp forest. Unfortunately, that fate remains largely unknown. Some of the kelp production is consumed by herbivores, most of which would be rapidly mineralized to carbon dioxide via respiration. The rest of this kelp carbon (probably the large majority) becomes detritus, which either is consumed and eventually respired by *detritivores* or sinks into the deep sea.

The detrital pathway creates a lag in the *mineralization* of kelp organic carbon, which is itself a storage effect. We do not yet know the size or temporal dynamics of the two main detrital storage compartments—that is, the detritivore pathway and the deep-sea-sinking pathway. This is a scientific frontier in need of exploration. If it turns out that these detrital storage compartments are small and/or short lived, then the storage effect of sea otters on atmospheric carbon is reasonably estimated by the amount of carbon held at any instant in time by living kelps. While that effect may be of local importance, the global effect is inconsequential. But if it turns out that the storage compartments are large and long lived, sea otters could indeed be meaningful players in controlling the global carbon cycle.

The preceding examples demonstrate that the otter–urchin–kelp trophic cascade penetrates the structure and dynamics of coastal ecosystems along numerous and serpentine pathways. But aside from these examples, most of the details of these complex, indirect interactions remain unstudied and thus unknown. I'll close this chapter with a bold prediction, which is that every species in the coastal zone is influenced in one way or another by the ecological effects of sea otters. I don't see how it could be otherwise. Eventually, I hope, time and effort will tell.

NINE

Sea Otters and the Red Queen

PLANT–HERBIVORE COEVOLUTION

IN LEWIS CARROLL'S *THROUGH THE LOOKING GLASS,* the Red Queen says to Alice, "Now, here [referring to Looking Glass Land], you see, it takes all the running you can do, to keep in the same place." More than 40 years ago, evolutionary biologists co-opted the "Red Queen" as a catchy metaphor for the process of coevolution between organisms that interact with one another agonistically (+/−; see chapter 2). A common example is the evolution of defense and resistance in consumer–prey systems. The idea is this: Consumers have negative effects on their prey; the prey seek to reduce those negative effects by evolving defenses; and the consumers, in turn, seek to overcome those defenses through the evolution of resistance. The outcome is an evolutionary "arms race" in which the antagonists seek to maintain or improve their respective places in nature. One might expect to see the outcomes of such evolutionary processes wherever the forces of ecology have been strong enough over sufficient periods to result in evolution through natural selection.

My first glimpse of this process in the sea otter–kelp forest system was motivated by Paul Dayton's simple but elegant experimental studies of competitive interactions in the kelp canopy at Amchitka Island, conducted in 1971–72 and published in 1975 (Dayton, 1975). Paul had recently completed his doctorate at the University of Washington and was headed for Scripps as a new assistant professor. While a graduate student, he had become acquainted with Phil Lebednick, another University of Washington student who was doing his graduate research at Amchitka. Phil had invited Paul to Amchitka to have a look at the kelp forests and perhaps even do some work of his own.

Upon arriving at Amchitka, Paul quickly saw a pattern in the distribution of kelp species with increasing water depth. Shallow areas, from the low-tide line to around 30 feet, were dominated by a dense canopy of kelp species in the genus *Laminaria*. At greater depths, *Laminaria* gave way to kelp in another genus *Agarum,* which continued to depths of about 60–80 feet, below which the kelp canopy dwindled to nothing.

Paul hypothesized that *Laminaria* was the competitive dominant in this system, thus excluding *Agarum* from shallower water. He began testing this hypothesis in 1971 by removing *Laminaria* from a swath of habitat that extended from just below the intertidal zone through the lower end of *Laminaria*'s depth range. Like any good experimental ecologist, he also established control swaths in which the *Laminaria* canopy was left intact. When Paul returned the following summer, he found that *Agarum* had indeed extended its depth range upward into shallow water in the *Laminaria*-removal swaths while the control swaths remained unchanged.

Paul's findings led me to wonder just what it was about *Laminara* that gave it a competitive advantage over *Agarum* in shallow water and why *Laminaria* didn't also exclude *Agarum* from deep water. I knew that the first part of this question would be hard to answer, but the second part seemed more manageable. One possible reason for the general absence of *Laminaria* in deep water was that it simply couldn't live there, perhaps because of reduced light or for some other reason; another possibility was that it was physically capable of living there but that increased water depth somehow reversed the competitive hierarchy between *Laminaria* and *Agarum*.

Agarum looked to me like an algal species that was made to live in deeper water, where water movement from wave surge is much less intense than it is in the shallower habitats where *Laminaria* prevailed. In contrast to *Laminaria,* which is robust, leathery, and strongly attached to the seafloor by a large holdfast, *Agarum* is delicate, flimsy, and only weakly attached to the seafloor by a small holdfast. I thus wondered whether *Agarum* wasn't somehow more efficient than *Laminaria* at making a living in the reduced light and lower wave energy of deeper-water habitats.

Paul Dayton's approach to the study of competitive interactions between *Laminaria* and *Agarum* and the results of his species-removal experiments strongly influenced me. I knew that *Laminaria* was capable of living in deep water because I had seen it there, albeit in limited abundance, at a few places around Amchitka. Together with Paul's findings, these observations led me to suspect that *Laminaria*'s absence from deeper water resulted from competition with *Agarum*. I therefore thought that I would simply extend Paul's experimental protocol into deep water to test that hypothesis, and I expected the results to provide both an interesting complement to Paul's findings and material for a chapter in my doctoral dissertation, which at the time I was struggling to define.

I began the study in the summer of 1972 by removing *Agarum* from a swath that extended downward from the lower reaches of the *Laminaria* canopy into deeper water. In the months that followed, I never saw the slightest hint of an effect of *Agarum* on *Laminaria*. Instead, *Laminaria* remained absent from the cleared swaths and *Agarum* recovered to its pre-removal distribution and abundance. Something other than competition for space or light was responsible for *Agarum*'s numerical dominance over *Laminaria* in deep water.

I never discussed these findings with Paul, in part because he had left Amchitka by the time I started the *Agarum* removal experiments, in part because I was humiliated by the lack of supporting evidence for the competition hypothesis, and in part because I couldn't think of any alternative hypotheses that seemed reasonable and were testable. As it turned out, Paul understood much of why *Agarum* replaced *Laminaria* in deep water, but I didn't know that because he hadn't mentioned it to me (or if he did, I missed it) and his paper wasn't published until 1975. In the interim, I figured it out for myself.

I knew at the time that sea otters ate urchins and that sea urchins ate kelp, but I didn't recognize what a destructive effect urchin grazing could have on kelp. I should have known this, because Bob Paine and Bob Vadas had shown that removal of sea urchins from tide pools on the coast of Washington caused fleshy macroalgae to proliferate (Paine and Vadas, 1969). But in truth, the potential destructive force of urchin grazing didn't occur to me until I saw it with my own eyes during our visit to Shemya Island in 1971 (see chapter 4). Not only was the effect of urchin grazing at Shemya apparent from the high urchin abundance and vast expanses of deforested seafloor, but urchins at Shemya occurred at their greatest abundance in shallow water, immediately below the lower reaches of the intertidal zone. The reason for the high urchin abundance in shallow water at Shemya was clear enough— there was little for them to eat in deeper water, and the accumulation of drift algae from the algae-dominated intertidal zone and by plants that occasionally washed ashore from elsewhere was greatest in these shallow habitats.

After seeing the patterns of urchin and algal distribution and abundance at Shemya and realizing that these were driven by the absence of sea otters, my understanding of the underlying reasons for a segregation of *Laminaria* and *Agarum* into their respective shallow and deep-water zones fell quickly into place. As I began measuring the co-occurring density and size of sea urchins, I saw that the depth-related patterns at Amchitka were reversed from those that occurred at Shemya. That is, only the occasional small urchin could be found in shallow water at Amchitka, while individual size and population density both increased with depth, which I concluded was a consequence of the increased cost of diving and, thus, the reduced foraging efficiency of sea otters in deeper water. As with all air-breathing vertebrates, the dive time of a foraging sea otter is limited by the animal's oxygen storage capacity, leading in turn to both increased travel costs and reduced amounts of time that can be spent on the bottom in search of prey in deeper water. That explained the sea otters' reduced foraging efficiency in deeper water.

These various observations and logic led to an alternative hypothesis for the absence of *Laminaria* and the abundance of *Agarum* in deeper water, which was that *Agarum* was somehow more resistant to sea urchin grazing. In Bob Vadas's (1970) study in the Strait of Juan de Fuca, urchins had shown a preference for *Laminaria* over *Agarum*. I repeated his work in a crude sort of way at Amchitka by translocating individual *Laminaria* and *Agarum* plants into deep-water urchin barrens and measuring the rates at which they were consumed by urchins. The urchins fed on both kelp species, but *Laminaria* was consumed at higher rates, lending support to the grazing-resistance hypothesis.

This discovery led to the question of how *Agarum* did this. I thought that it must have something to do with a difference between the species in nutritional quality. I had no idea what that might be, but there were numerous possibilities. Perhaps the energy density of *Laminaria* was greater than that of *Agarum*, thus making *Laminaria* more palatable to sea urchins. Or it might be that *Agarum* produced some kind of *secondary metabolite* that deterred urchin grazing. The appeal of the secondary-metabolite hypothesis was that it also explained why *Laminaria* enjoyed a competitive advantage over *Agarum* in shallow water. The idea was simply this: *Laminaria*'s competitive advantage over *Agarum* in shallow water arose from not having to invest photosynthetic resources into the biosynthesis of defensive secondary metabolites. *Agarum* thus gained the advantage in deep water because these (imagined) secondary metabolites protected it from the higher intensities of herbivory that occurred there.

The sticking point in the early 1970s was that this was only an idea. I knew from Dayton's work that *Laminaria* was the competitive dominant, and

I knew from my own studies and earlier work that *Agarum* was low in the hierarchy of food preferences by sea urchins (Vadas, 1970). I suspected that secondary metabolites were at the center of these differences, but I didn't know what those metabolites might be, much less whether they differed in concentration between *Laminaria* and *Agarum*. I wasn't even sure that the difference between *Laminaria* and *Agarum* had anything to do with secondary metabolites.

Plant secondary metabolites had been known to deter terrestrial herbivores for some time, but there was little knowledge of the function of such compounds in the marine realm. That all began to change in the mid- to late 1970s with the beginnings of collaborative studies between marine natural-products chemists and reef ecologists. I was especially influenced by the early work of Mark Hay, whose doctoral research on coral reef plants in the Caribbean led him to suspect that plant secondary metabolites played important roles in determining the distribution and abundance of coral reef algae through their deterrent effects on herbivores. Mark eventually joined forces with Bill Fenical, a natural-products chemist who could identify these compounds and quantify their concentrations in the algae. Working together, Mark and Bill helped pioneer a new subdiscipline of marine chemical ecology, repeatedly showing clear effects of various plant secondary compounds in modulating the outcomes of consumer–prey interactions.

Hay and Fenical worked mostly in the tropics and mostly on red and green algae. I was interested in brown algae and cold-water ecosystems. Although I suspected similar controlling processes between these groups of species and ecosystems, I was also quite certain that the details would differ. I needed a collaborator—someone who was interested in exploring the chemical ecology of cold-water algae. I found one in about 1980 when I met Peter Steinberg (figure 9.1), who had received his bachelor's degree from the University of Maryland and had come to Santa Cruz to study chemical ecology.

Peter was interested in kelp and other brown algae, especially how these plants produced and used secondary metabolites to defend themselves against herbivores. His interest was focused on a group of compounds known as *phlorotannins,* large molecules that were ubiquitous in the brown algae, varied considerably in concentration across different algal species, and, at least in some cases, appeared to deter herbivores. Natural-products chemists had avoided working on phlorotannins because of their large size and extreme structural complexity. I also worried about this, especially after Hay and Fenical demonstrated that minor alterations in the chemical structure of

FIGURE 9.1. Frank "Pancho" Winter (left) and Peter D. Steinberg during an algae-collecting trip to New Zealand's South Island. (Photo by author)

the much simpler compounds they were studying could have large effects on their deterrent qualities. But Peter and others had not found evidence of such effects in the phlorotannins. It was beginning to look like deterrence by phlorotannins was dose dependent and largely independent of the algal species from which they were derived. Moreover, phlorotannins were easy to work with because of their structural stability and the relative ease with which they could be extracted from algae and assayed.

Shortly after meeting Peter, I began to wonder whether the different distributions and competitive abilities of *Laminaria* and *Agarum* were related to differences in their phlorotannin concentrations. The biosynthesis of elevated phlorotannins might defend these plants against the increased intensity of deep-water herbivory and, because of the presumed cost of making these compounds, explain *Agarum*'s inability to compete with *Laminaria* in shallow water. I was excited by this hypothesis because it provided a potentially more mechanistic understanding of the patterns I had seen in my field studies and in both Paul Dayton's and my own experimental results.

The first step in testing the hypothesis was relatively straightforward. We would measure phlorotannin concentrations in *Laminaria* and *Agarum* to determine whether they were indeed greater in the latter. My initial concern was about getting plant specimens home from the Aleutians with their secondary metabolites intact. Peter had confronted this issue before and assured me that frozen material would likely work just fine for our crude initial analyses. So the following summer, I collected a small number of individuals of the common kelp species in the Aleutians and shipped them frozen back to Peter at UC Santa Cruz. It wasn't long before the results were in. Phlorotannin concentrations in *Agarum* were among the highest Peter had measured in marine brown algae (about 4 percent dry mass), whereas those in *Laminaria* were about an order of magnitude less. This was exciting. We seemed to be on the right track.

This line of inquiry—from the initial observation that algal species composition and sea urchin abundance varied with depth at Amchitka Island; to Dayton's experiments on algal competitive interactions; to Vadas's studies of sea urchin feeding preferences; to my own work on algal competitive interactions and the susceptibility of algal species to grazing damage from sea urchins; to the associated differences in phlorotannin concentration between the same algal species—enriched my understanding of the otter–kelp trophic cascade and led to the beginnings of an evolutionary hypothesis for how this richness in community structure and function came to be. My hypothesis was this: The sea otter's ability, as an air-breathing predator, to limit urchin populations declined with water depth. This physiological and behavioral constraint on sea otter foraging efficiency led to an increase with water depth in the size and abundance of sea urchins, and therefore in the intensity of herbivory. The depth-related gradient in herbivory, in turn, influenced the evolution of kelp by selecting for a competitively superior shallow-water flora and a defensively superior deep-water flora. The divergence of these depth-related life history strategies in the kelps was driven by the costs and benefits of phlorotannin biosynthesis. Peter and I published this idea and its supporting evidence a few years later (Estes and Steinberg, 1988).

The hypothesis made for a good story but, like so many macroevolutionary hypotheses, was difficult to test. I knew that many in the always skeptical scientific community would view Peter's and my hypothesis as a just-so story—that is, an explanation made up to fit the observed patterns, with no rigorous testing of the hypothesis or possible alternatives. And there were alternatives that deserved consideration. Our hypothesis rested on the key assumption that phlorotannin biosynthesis came at a metabolic and material cost to the algae. The difficulty was in actually demonstrating such a cost. Although I find it difficult to imagine the absence of a cost in making phlorotannins, a cost still hasn't been rigorously demonstrated (Dworjanyn et al., 2006). Moreover, we could not be certain that increased levels of deep-water herbivory led to the evolution of elevated phlorotannin concentrations in *Agarum*. This might have happened for some different reason, thus predisposing *Agarum* to do well in deeper water.

These difficulties led Peter and me to think more expansively about how the sea otter–kelp evolution hypothesis might be independently tested. Our story at the time was an explanation, after the fact, for what we had seen in the Aleutians. We pondered the question of how we might evaluate the hypothesis more rigorously by formulating a prediction and then testing that prediction with observations and experiments. We recognized that the question would not be resolved by thinking small, which is to say that interesting answers were unlikely to be forthcoming from continuing to look at the North Pacific Ocean in further detail. After all, whatever sea otters might have caused in the way of evolutionary responses to species at the temperate margins of the North Pacific Ocean would likely have occurred everywhere across that region. Even if that presumption weren't entirely true, we had no clear sense of where to look or exactly what to look for. We needed another kelp forest ecosystem, one that had evolved independently and in the absence of sea otters or otter-like ecological analogues.

Once our thinking had developed to this point, a path forward became clear. We would look at kelp forests in the Southern Hemisphere. The underlying logic was simple enough, at least in broad brush. Temperate to subantarctic coastal ecosystems in the Southern Hemisphere were physically similar to those at similar latitudes in the Northern Hemisphere. Temperate reefs in both hemispheres contained the same key functional groups (i.e., fleshy brown algae, grazing sea urchins, and grazing gastropods), and the species in these functional groups in the two hemispheres had evolved in isolation of one another because of the tropical barrier to dispersal that lay between them. We therefore imagined that while kelp forest ecosystems in the two hemispheres might appear to be similar, most of the component species were only distantly related because of the long period of isolation and independent evolution. We knew that the presence or absence of sea otters and their recent ancestors wasn't the only difference between the hemispheres' kelp forest ecosystems. But we also reasoned that the selective influence of the sea otter's lineage in the Northern Hemisphere must have been very strong and, therefore, that an evolutionary response to that selective influence ought to be apparent between Northern Hemisphere and Southern Hemisphere kelp forests.

The assumption that Northern and Southern Hemisphere reef systems evolved under differing intensities of predation was fundamental to our hypothesis and was seemingly borne out by the available literature. In fact, Southern Hemisphere herbivores are subject to predation. We knew this from the published literature. But the predators are mostly fish and crustaceans, neither of which appeared to be as effective as sea otters in limiting their prey populations. So far as we could determine, Southern Hemisphere kelp forest ecosystems had never included a *homeothermic,* air-breathing predator like the sea otter.

As our interest in contrasting Northern and Southern Hemisphere kelp forests intensified, the specific questions to ask and the methods of answering those questions began to come into focus. But before venturing down those paths, we needed to carefully review what was known about the climatological and *phylogeographic* histories of the world's temperate oceans. The first step in this process was to establish a view of the broad scales of space and time over which our comparisons would be made. The time scale was clear enough—we were interested in the current glacial age, or the period beginning in the late *Miocene* to early *Pliocene* (roughly from 5 million to 3 million years ago). Subtropical to tropical conditions prevailed at high latitudes before that time, and thus modern, cool-adapted biotas of shallow, temperate to subarctic seas had not yet begun to evolve before the late Miocene.

A spatial template for making *biogeographic* contrasts was also required. For this purpose we imagined the world's cold coastal oceans as being divisible into five (or six) distinct regions: the North Pacific; the North Atlantic; southern South America; southern Africa; and southern Australia, Tasmania, and New Zealand. We had neither time nor resources to look at each of these regions in detail. We would pick just one to contrast with the North Pacific, at least for the initial study. We immediately discarded the North Atlantic because it was too closely connected to the North Pacific by way of the numerous *transarctic interchanges* that occurred across the Arctic during the Pleistocene. North Pacific and Atlantic kelps and sea urchins, in particular, are closely related. South America might have proved interesting, but we rejected it as a primary study location because the South American brown algal flora is highly impoverished. We further rejected southern Africa because it appeared to us as though the area had recently been invaded by kelps from the Northern Hemisphere. In particular, *Laminaria* were reportedly spreading in southern Africa. Closely related species of *Laminaria* also occurred in the cool, deep, extremely clear waters of eastern tropical South America, and we therefore suspected that these predominantly Northern Hemisphere kelps had managed to reach southern Africa from a Northern Hemisphere center of origin by recently dispersing beneath the tropical warm-water barrier. Whether or not that was the case, the presence of *Laminaria* in southern Africa was problematic to the contrast we wished to make, which required biotas that had evolved in isolation from one another. Southern Australia and New Zealand seemed like the better choice because those regions supported diverse brown algal floras and grazing macroinvertebrate faunas that appeared to have evolved largely in geographic isolation, and thus independently from those of the North Pacific.

The next question was what to measure. Our hypothesis held that differing food-chain lengths between the North Pacific on one hand (a three-tiered food chain) and temperate Australasia on the other (a two-tiered food chain) ought to have led to fundamentally different coevolutionary dynamics between the algae and their herbivores (figure 9.2). We hypothesized that the New Zealand flora would be comparatively well defended, much like the deep-water system at Amchitka, because of the chronically high intensities of herbivory that purportedly occurred there over at least the past several

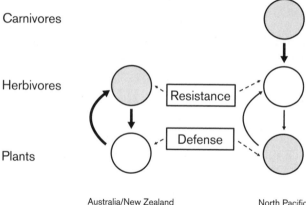

Carnivores

Herbivores

Resistance

Defense

Plants

Australia/New Zealand   North Pacific

FIGURE 9.2. Model depicting the influence of food-chain length on the coevolution of defense and resistance in plants and their herbivores. Gray shading indicates resource limitation; lack of shading indicates consumer limitation; thickness of solid arrow-lines indicates interaction strength. In two-trophic-level systems (left; such as Southern Hemisphere kelp forests), the limiting influence of herbivores on plants is strong and thus leads to the coevolution of defense and resistance. The addition of a third trophic level (right; such as Northern Hemisphere kelp forests with sea otters) reduces herbivore abundance, thus reducing the intensity of herbivory on plants and eliminating the coevolutionary arms race between plant defense and herbivore resistance.

million years. And we suspected, moreover, that those defenses would be realized through high concentrations of phlorotannins. If this were true, we imagined further that Southern Hemisphere herbivores would have evolved superior abilities to resist those defenses. At the landscape level, we hypothesized that the rates of herbivory in Australasia would fall somewhere between those we had measured in Northern Hemisphere systems with and without sea otters, and that brown algae and their herbivores in the Southern Hemisphere would live in a more intimate association with one another as well. All that remained was to go to Australia or New Zealand and have a look.

After completing his doctorate, Peter moved to Sydney on a Fulbright postdoctoral fellowship with the intent of beginning the survey. His plan was to familiarize himself with the regional biota, collect as many species from the brown algal flora as possible, and determine their phlorotannin concentrations. He did this and found that virtually all of the common plants were chock-a-block full of phlorotannins (Steinberg, 1989). Peter subsequently found that phlorotannins were not especially deterrent to Australian herbivores. These findings were terrifically exciting because they supported our hypothesis. Armed with that knowledge and information, Peter and I were able to obtain support from the National Science Foundation for a more in-depth study, one that would be done primarily in New Zealand.

We chose New Zealand because Howard Choat and Bob Creese, both faculty members at the University of Auckland, were interested in the study and therefore arranged the necessary logistical support for our project at the university's Leigh Marine Laboratory. Leigh Marine Lab, on the eastern shore of New Zealand's North Island and about a two-hour drive north of Auckland, was an ideal base of operations for our work on Australasian kelp forests. It offered logistical support for diving, access to the Leigh Marine Reserve as a place for fieldwork that was largely free from human exploitation and other disturbances, and a seawater system that would be needed for some of our planned experiments. The Leigh Lab also provided access to the analytical equipment we would need for measuring phlorotannin concentrations in algae and for extracting and purifying these compounds. Finally, the Leigh Lab was buzzing with students who were eager to share their knowledge of local natural history and help with the work.

Our first goal in New Zealand was to survey the algal chemistry, which we did by traveling around the country, collecting as many species of brown algae as we could find, and measuring their phlorotannin concentrations. There were no surprises here. The Australian and New Zealand floras were very similar in both species composition and phlorotannin chemistry. Phlorotannin concentrations in these floras ranged upward to almost 16 percent of the plant's dry mass, with an average across species of about 10 percent. All the common species were rich in phlorotannins. By distinct contrast, phlorotannin concentrations in brown algae of the North Pacific were mostly less than 5 percent, with an average across species of just over 1 percent dry mass. The common North Pacific species all were poor in phlorotannins.

The next step was to determine whether North Pacific and Australasian herbivores differed in their ability to resist phlorotannins. Initially we thought we might do this by merely looking at the rates at which these two suites of herbivores consumed algal species with differing phlorotannin concentrations. But that approach was problematic for several reasons. For one, the different algal species varied in numerous ways besides phlorotannin concentration, so we could not be certain that differences in grazing rates among these species were responses to phlorotannin concentrations. For another, we would not know the exact phlorotannin concentration in the experimental algal materials because the concentration varied to a small degree among individuals within species and across different tissues of the individual algae. And finally, for logistical and legal reasons, we could not expose the herbivores from one ocean basin to algae from the other, something we very much wanted to do in order to determine whether there was some qualitative difference in phlorotannin chemistry between the two oceans that might influence their deterrent effects on herbivores.

Because of these various difficulties, we settled on a different approach—one in which we would add phlorotannins in varying concentrations and from various algal species to a common and otherwise neutral grazing model. That model was freeze-dried *Ulva* embedded in agar disks. *Ulva* is a widely distributed intertidal green alga that is readily eaten by a broad suite of marine herbivores. Moreover, it does not contain measurable concentrations of phlorotannins. We discovered that agar disks containing freeze-dried and then ground-up *Ulva* were readily consumed by all the North Pacific and Australasian herbivores we tested, and that these herbivores survived and grew in the laboratory on an exclusive diet of the *Ulva*-containing disks.

Fortified with this knowledge, it was a simple matter of adding purified phlorotannins from the various algal species at whatever concentrations we wished to the *Ulva*-containing disks and determining whether these added compounds deterred grazing by the various herbivores. We did this by offering the herbivores disks with and without phlorotannins and measuring the amount of each that was consumed over a 24-hour period. We added phlorotannins at equivalents of 5 percent and 13.4 percent dry mass to mimic the lower and upper range in concentrations of phlorotannin-rich algae. We were also able to transport the extracted and purified phlorotannins from North Pacific and Australasian algae between hemispheres, thus exposing both North Pacific and Australasian herbivores to phlorotannins from a common group of algal species that represented both ocean basins.

The results of this interhemispheric herbivore-feeding experiment fell closely in line with our initial predictions. North Pacific herbivores were strongly deterred by phlorotannins, regardless of source or concentration, whereas Australian herbivores were either undeterred or less strongly deterred in the same test (Steinberg et al., 1995).

Armed with an understanding of differences in algal defense and herbivore resistance between the North Pacific and Australasia, we next returned to the field to examine the consequences. We had already measured the rates

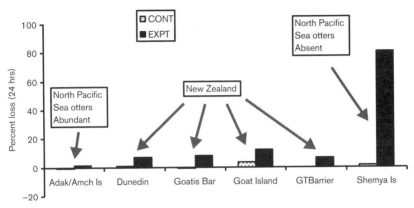

FIGURE 9.3. In situ rates of herbivory in New Zealand, and at sites with and without sea otters in the North Pacific Ocean (from Steinberg et al., 1995).

of herbivory and the distribution and abundance of algae and their major herbivores on North Pacific reefs with and without sea otters (see chapter 7). What remained was to repeat these measures in Australasia.

The grazing assays in New Zealand were conducted just as in the North Pacific, except that we used the common local algal species to measure rates of tissue loss. Our fieldwork in New Zealand was done in marine protected areas to minimize any possible effects of human take on the nature and strength of ecological interactions.

The overall findings were interesting and very informative. As described above, we placed preweighed pieces of algal tissue from four common local algal species onto a reef and measured the tissue loss to herbivory by comparing uncaged treatments with caged controls after a 24-hour period. We saw no appreciable change in mass in any of the caged controls in New Zealand or in areas of the North Pacific with or without sea otters. These findings told us that the caged controls were equally effective at excluding herbivores in each of the three major settings. The more telling result, however, was the pattern of algal tissue loss in the uncaged treatments. Where sea otters were abundant in the North Pacific, algal tissue loss to herbivory was essentially nil. Where sea otters were absent in the North Pacific, algal tissue loss to herbivory was high—about 80 percent. And in New Zealand, about 7 percent of the algal tissues were lost to herbivory—a rate that was significantly higher than that in the caged controls, significantly higher than at North Pacific sites with sea otters present, but significantly lower than in the North Pacific where sea otters were absent (figure 9.3).

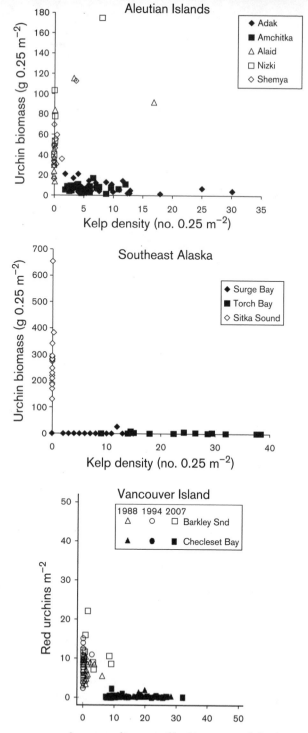

FIGURE 9.4. *State space* diagrams of herbivore versus kelp abundance in the North and South Pacific oceans, expressed as number (no.) or grams (g) per quadrat (0.25 m^{-2}) or as number per square meter (m^{-2}) (from Estes et al., 2013).

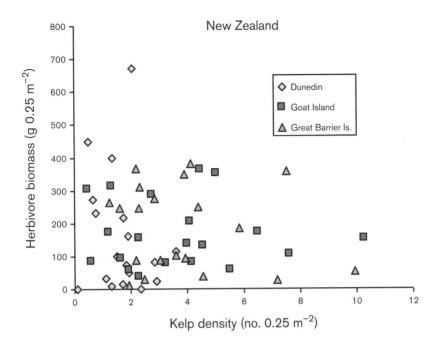

The spatial distribution of algae and herbivores also differed markedly between North Pacific and Australasian reefs. These two groups of organisms almost never co-occurred in any abundance in the North Pacific. In Australasia, however, the algae and their herbivores commonly lived in intimate association with one another (figure 9.4).

What did all of this tell us about sea otters and the Red Queen hypothesis? To the casual eye, North Pacific and Australasian reefs look fairly similar to one another, in that both are commonly covered with algae. But with closer examination, one sees that these systems have been assembled in different ways. Australasian algae and their herbivores appear to be products of an arms race in which the algae have evolved purportedly costly defenses to the chronically high rates of tissue loss to herbivory, and the herbivores in turn have evolved resistances to those defenses (figure 9.2). This coevolutionary process must have progressed in Australasia in intimate lockstep over the millennia, driven by the fundamentally two-tiered structure of the kelp forest food web. The result was an ecosystem in which the algae's competitive abilities have been compromised by the need to defend themselves against their herbivores, and the herbivores' foraging rates and efficiencies have been compromised by the need to deal with plant defenses. An interesting consequence of these compromises is the evolution of a system in which the algae and the herbivores are able to live together.

Were it not for sea otters and their recent predecessors, the same patterns might have evolved in the North Pacific. But these predators did arise in the North Pacific several million years ago, thereby adding a third trophic level to the kelp forest food web, which in turn broke the potential for an evolutionary arms race between the algae and their herbivores. Without the need to defend themselves from chronic herbivory, the kelps invested their resources in growth and competitive ability. Without the need to resist those defenses, the herbivores evolved a capacity to eat as much as possible and to transform these food resources into replacements for the large losses to sea otter predation. The North Pacific system was not held together by the coevolution of defense and resistance in the algae and the herbivores, but by a predator that prevented the herbivores from consuming very much of the algae. This lack of coevolution in defense and resistance explains why North Pacific algae and their herbivores are unable to live together, and why urchins so often destroy North Pacific kelp forests when their predators are lost from coastal ecosystems.

After developing this differing view of algal–herbivore coevolution between the North Pacific and Australasia, I began to wonder how the regional floras' distinctly different nutritional profiles had affected other algal consumers. The possibilities struck me as being nearly endless. I wondered whether some enterprising young scientist might not build an exciting career around the issue, but so far none has. My own career had already been defined by a wider range of issues that revolved around the otter–urchin–kelp trophic cascade. So I chose to venture only briefly down this path, by thinking about and looking at two groups of herbivores—the hydrodamaline sirenians (which culminated in Steller's sea cow) and the haliotid gastropods (abalones).

Hydrodamaline sirenians radiated into the North Pacific from a common ancestor with modern dugongs, a tropical seagrass consumer. As the poles cooled and algal forests developed at higher latitudes, the hydrodamalines evolved as algivorous marine mammals (Domning, 1978). Exactly how they managed to flourish on an algal diet is unknown, because the last sea cow perished in 1768. However, their transition from tropical seagrass feeder to cold-water algal feeder was accompanied by greatly increased body size (adults may have reached 30 feet in length) and the replacement of teeth for grinding the abrasive seagrasses with a palatine plate for mashing the much softer and nonabrasive algae.

The sea cow's early dugong-like ancestor was pantropically distributed, which means they might have become algal feeders by radiating poleward in any ocean, or even in multiple cold-water oceans. But this only happened once, an event that led them into the North Pacific. The fact that they wound up in the North Pacific may be a product of chance alone. If we could somehow rerun the clock of Earth history, we might find that sea cow–like creatures evolved as consumers of algal floras that occupied the North Atlantic, South America, southern Africa, or Australia–New Zealand. It could be that such creatures did evolve but the fossil remains of these extinct, algal-eating sea cows remain undiscovered. Or it could be that the North Pacific offered the only algal flora of sufficient nutritional quality to allow for the evolution of a cold-water-living mammalian *algivore*. Whether the nutritional profiles of the other cold oceans' algal floras prevented the evolution of an algivorous marine mammal is largely a matter of conjecture. But the great algal abundance and the lack of inhibitory or toxic compounds in North Pacific algae, it seems to me, could not but have helped draw the hydrodamaline sirenians into the North Pacific Ocean.

My interest in abalones arose through a long-standing friendship with David Lindberg, a former Santa Cruz student who moved to UC Berkeley after completing his doctorate to curate the invertebrate collection in the university's Museum of Paleontology. Among David's many intellectual attributes was an in-depth knowledge of mollusks. My understanding of historical biology was enriched on numerous occasions over the years by one fact or another that appeared almost magically from David's tremendous repertoire of knowledge. I don't know what it was that got him thinking about abalones, but he recognized that the information he was compiling on the distributions and maximum body sizes of living and fossil abalones was germane to my thinking about kelp evolution. Even before he rigorously compiled the data, David had a sense of the patterns and why they were interesting. This began with something that many people knew but no one had sought to explain, which is that the North Pacific Ocean supported the world's most diverse and largest abalones. The king of these molluscan giants is the red abalone *(Haliotis rufescens)*, which on occasion grows to more than a foot in shell length.

The abalones are an old, diverse, and widely distributed group. Numerous extinct species are known from the fossil record, the oldest fossil abalones appearing in the Late Cretaceous, more than 60 million years ago. And except for the North Atlantic, which has always lacked an abalone fauna, these animals are worldwide in occurrence from the tropics to the temperate zones.

The first question that David and I asked was whether modern warm- and cool-water abalones differed in maximum body size. The answer was clear: All tropical abalones are small, less than 10 centimeters in maximum shell length, whereas temperate regions of the world consistently support faunas with much larger species (Estes et al., 2005).

Abalones are herbivores, and we suspected that the size difference between tropical and temperate species had resulted from the comparatively much greater algal abundance in cool seas compared with the tropics. We couldn't demonstrate the proposed mechanism, but we could look back in time to determine whether abalones have always been large or, if not, when they became so. Here again, the pattern was clear. From the Miocene and before, all of the abalones were small—similar in maximum body size to modern tropical species. But beginning in the Pliocene, which marks the onset of polar cooling and thus the presumed development of algal stands at temperate latitudes, the maximum body size of abalones became much larger. This discovery was consistent with our developing thesis that the evolution of large body size in temperate abalones was driven by the high productivity and abundance of kelp and other macroalgae.

Our next question had to do with the manner in which large abalones evolved from their smaller ancestors. We wondered, in particular, if the body-size transition occurred once or multiple times. If it occurred just once, then the body-size patterns in fossil and living abalones might be little more than an accident. But if it occurred more than once, then selection for large body size in temperate environments could be invoked with greater confidence. David brought in a colleague, Charlie Wray, to sequence the mitrochondrial DNA of as many extant abalones as he could obtain samples from. David then assembled the various species into a *phylogenetic tree,* onto which he mapped maximum shell length. This analysis established that large body size had independently evolved at least twice in the abalones.

We knew that the largest of all abalones occurred in the North Pacific Ocean and that the algal flora of this region was low in phlorotannin concentrations, but we didn't know whether phlorotannins inhibited growth or maximum body size in North Pacific abalones. Frank (Pancho) Winter (figure 9.1), a graduate student in my lab, answered that question by growing recently settled red abalones in the laboratory on both algae and agar disks with high or low phlorotannin concentrations. Pancho demonstrated a clear inhibitory effect of phlorotannins on abalone growth (Winter and Estes, 1992).

To my way of thinking, the sea cow and abalone stories were mutually reinforcing in that each suggested evolutionary responses to a North Pacific brown algal flora that was low in inhibitory secondary metabolites and therefore of high nutritional quality. For the abalones, which could be observed and studied in modern ecosystems, I was able to step back and put everything we had learned into an integrated perspective. I saw abalones as an ancient group of organisms that, throughout much of their evolutionary history, remained small, just as modern abalones are small in tropical–subtropical oceans. With late Cenozoic polar cooling and the associated development of kelp forests at higher latitudes, abalones grew larger in response to an increased abundance of their macroalgal food, which fueled their growth through detrital fallout from the overlying kelp forest canopy.

This pattern of increased body size in cool-water abalone faunas occurred across the world's oceans, but it was most pronounced in the North Pacific, where the abalones' algal food was most nutritious. Although large abalones were vulnerable to predation by the evolving sea otter lineage, so were the smaller ones. Size in abalones conveys no advantage or disadvantage with regard to sea otter predation. The North Pacific abalones were able to grow to large sizes because they survived as sedentary drift-algal feeders in the rocky seafloor's cracks and crevices. While predation by sea otters on herbivorous macroinvertebrates in the more exposed habitats led to the evolution of a nutritious algal flora, North Pacific abalones, safe from otter predation within their cryptic habitats, became large and thrived because of the abundance of high-quality detritus that drifted down to them from the overlying kelp canopy.

These evolutionary stories will seem to some to be just that—ideas that have not been tested as rigorously as they should be. I'll be the first to admit that alternatives may well exist that have not been properly explored and discounted, if indeed they are even discountable. But until someone comes up with a better story, the one I have told rings true to me because it holds together so well through its internal and external consistencies. To my mind, the underlying logic spins off the Red Queen hypothesis with simple elegance: The radiation of sea otters into the North Pacific broke an evolutionary arms race between macroalgae and their herbivores and thereby created a productive and highly nutritious flora, which in turn fueled the evolution of North Pacific herbivores and detritivores. Manifestations of this evolutionary scenario should be widespread across the world's oceans. All that remains is for someone to look and determine whether that is true.

TEN

Sea Otters and Killer Whales

MY FIELD PROGRAM IN THE ALEUTIAN ISLANDS had begun to stagnate by the late 1980s, or so it seemed. I had been following the recovery of sea otters on Attu Island for almost 15 years and, except for the population's steady increase in numbers and range and the associated disappearance of large sea urchins, there had been little change in kelp abundance. I thought that the shift to a kelp-dominated state would eventually occur, but, having no idea when that might be, I was finding it increasingly difficult to justify the time and money required to simply continue watching. That Attu might remain an urchin-dominated ecosystem in perpetuity lingered in my mind as a troubling but important possibility, although I recognized that much more time and money would be needed to properly chronicle that eventuality. I had documented large-scale patterns of generality and variation in the otter–urchin–kelp trophic cascade in the Aleutian Islands by extensively sampling multiple islands with and without sea otters (see chapter 7). More effort, I knew, would add little to an already well-chronicled story. Any significantly expanded view of generality and variation in the otter–urchin–kelp trophic cascade would only come from looking elsewhere, in places such as southeast Alaska, British Columbia, or Russia. A second module of conceptual interest in the otter–urchin–kelp trophic cascade—its knock-on effects on other species and ecological processes—offered an almost infinite number of possibilities for future study. But we had completed two of these— the kelp production work with David Duggins and Si Simenstad, and the gull foraging work with David Irons and Bob Anthony—which was enough to establish the overall effect (see chapter 8). I was sure that other such knock-on effects occurred in the system. The challenge was not so much in discovery of the unknown but in devising ways to chronicle the obvious. These efforts, while important, were becoming more tedious than exciting. My interests at the time had also turned toward what I had long thought would be the third and final conceptual module of my research program on sea otters and kelp forests: plant–herbivore coevolution (see chapter 9). For that reason, I had begun spending long periods of each year in New Zealand. Much of what I had dreamed of doing as a nascent scientist in the early to mid-1970s was done, or was close to done, by the late 1980s. Although it had been an exciting and productive 15 years, it seemed that my time in the Aleutian Islands might be drawing to a close.

But fate intervened. I was in San Diego for a meeting—I can't recall what it was about, or even the exact year. Don Siniff, a well-known member of the marine-mammal research community, was there from the University of Minnesota. Don had recently begun working on sea otters in central California and Prince William Sound. By chance we had dinner together one evening, and talk inevitably turned to sea otters and what more of interest and value might be done with them. Don had been thinking about the Aleutians, and especially about Amchitka Island, where the sea otter population was apparently at or near carrying capacity and had been for decades. He was intrigued by Amchitka because it was one of the few remaining places in the world where a mammalian carnivore lived in the absence of any obvious human disturbances. The sea otters at Amchitka had been heavily exploited during the maritime fur trade, but that had ended almost a century earlier and the population had long since recovered. There were other such islands, but Amchitka provided both a chronicle of the sea otter's population status since the early 1940s and the infrastructure needed to support a remote field-research program. In Don's mind, and in mine as I listened to him, Amchitka offered far more than a place to describe an otter population that was at or near carrying capacity. It offered a rare opportunity to characterize the *demography* and behavior of a large carnivorous mammal living under the condition of resource limitation. That opportunity was made all the more appealing by the existence of nearby islands that only recently had become recolonized by sea otters and therefore supported populations that were well below carrying capacity. We imagined using interisland comparisons to answer yet another interesting ecological question, in this case how behavior and demography varied as a function of population density. Don thought that the National Science Foundation might be interested in supporting the work and suggested that we write a proposal to "test the water."

From our collective experience with sea otters and my intimate knowledge of Amchitka Island, we were easily able to identify an interesting set of hypotheses and how to go about testing them. Don and his colleagues had pioneered a radiotelemetry system for tracking otters that enabled a detailed look into the lives of individual animals. We would use this approach to contrast key elements of the sea otter's life history and behavior between islands of varying population status. We would measure age-specific fertility and mortality; body condition; diet; and activity budgets, especially time spent foraging. The National Science Foundation liked the idea and decided to support our proposal.

In the summer of 1990 we set off for Amchitka on the *Alpha Helix* with the immediate goal of capturing and putting radio transmitters on 100 sea otters. Two decades earlier, as part of an effort to reestablish this species in Oregon, the Alaska Department of Fish and Game had captured more than 100 otters from Amchitka in less than a week by deploying floating gill nets in or near surface kelp canopies. The unsuspecting otters, apparently thinking they were swimming through strands of kelp, became entangled in the nets, after which they were removed to holding boxes. We would use the same capture method, supplementing the net-catches with a more recently developed technique in which resting animals were seized from below by divers using rebreathers, underwater scooters, and a basket-shaped trap affixed to the front of the scooter (Ames et al., 1983). Although removing otters from the floating nets could be an adventure, both methods had proved safe and effective.

The captured otters were to be transported to the ship, where they would be sedated, examined, measured, and equipped with a radio transmitter that was surgically implanted by a wildlife veterinarian (Williams et al., 1981). The sedation was then reversed, after which the animal was placed in a holding pen for observation and then released back to the wild. This method had been used extensively in California and elsewhere in Alaska. Risk to the otters was low, and the transmitter-equipped otters were easy to relocate with a radio receiver. I hardly doubted our ability to capture and put transmitters on 100 otters in the almost three weeks I had allotted to this phase of the project.

But enigmatic results tormented me from the project's outset. Despite a stretch of excellent weather, the capture of 100 animals proved to be far more difficult than I had expected. During the Alaska Fish and Game operation in 1971, we had obtained as many as 15 otters in a single net set of two to three hours. In 1991, I was lucky to get one or two animals from the same effort at the same locations. The frequency of unsuccessful capture attempts increased as the days passed, and after almost three weeks of intense effort my team and I had managed to capture and attach transmitters to only 90 animals. That was enough for the purpose of the study, but I was perplexed by our inability to obtain more. I wrote it off to fate, thinking that perhaps the weather was too good and that the animals avoided our nets for that reason. Be that as it may, I finished the project's initial capture phase with a feeling of success and optimism for the forthcoming data-gathering phase.

But here again, the results were out of line with my expectations. While working at Amchitka in the early 1970s, I had found dozens of sea otter carcasses or their remains during daily walks along the beaches, the presumed consequence of density-dependent starvation in a population that was at or near carrying capacity. But similar beach searches in the 1990s yielded little more than the bleached bones of otters that had died years earlier. Our records from the tagged animals showed that the birth rate, which tends to be largely invariant among sea otter populations, hadn't changed. These births should have been balanced by deaths, and it was mystifying that the presumed surplus in production beyond carrying capacity wasn't appearing on the beach as it had in the 1970s.

The absence of stranded carcasses wasn't the only thing that had me scratching my head. The otters I had studied at Amchitka in the early 1970s fed extensively on kelp forest fishes, but we saw none of that in the 1990s. I wondered why the otters had stopped eating the fish. The otters' activity budgets also didn't match my expectations. Otters at Amchitka spent more than 60 percent of the daylight hours foraging in the 1970s, but our data from the early 1990s yielded estimates in the general range of 35–50 percent (Gelatt et al., 2002). This difference could have been due to a methodological bias, because different methods were used in the two studies—visual *scan samples* in the 1970s and telemetry-based data from focal animals in the 1990s—but I doubted that explanation. If anything, the scan-sampling data should have provided a lower overall estimate of the percentage of time spent foraging, because animals submerged in foraging dives were more likely to be missed in instantaneous scan samples than in telemetry-based focal follows. The most reasonable interpretation of these data was that the animals were now spending less time foraging.

And then there was another surprise. During the first winter, we began to observe otters feeding on smooth lumpsuckers (Watt et al., 2000), an oceanic fish species that periodically migrates into coastal waters to spawn and die.

Karl Kenyon had observed sea otters eating lumpsuckers at Amchitka in the 1960s, but I had never seen it anywhere in the Aleutians during the 1970s and '80s. Now, in the early 1990s, lumpsuckers were everywhere, and in tremendous numbers. We saw them around us while diving. And in the spring their carcasses littered the beaches in windrows across the central and western Aleutian archipelago.

The lumpsuckers provided a pulsed (periodic) nutritional subsidy to the otters during winter and spring, a time when many otters would otherwise be starving. Initially, that seemed to explain much of what we were seeing in the 1990s, including the absence of stranded sea otters on the beaches, the otters' improved body condition (Monson et al., 2000), and the reduced amounts of time they spent foraging (Gelatt et al., 2002). It wasn't a process I had gone back to Amchitka to explore in the early 1990s, but it was an interesting discovery nonetheless. As we finished the fieldwork at Amchitka several years later, I saw little reason to pursue the project's planned second phase of working at a nearby island where otters were below carrying capacity. Because of the lumpsuckers, I thought that was what we had just done at Amchitka.

But unanswered questions remained. Although the lumpsucker hypothesis predicted no difference in sea otter population size between the 1970s and '90s, accumulating evidence suggested otherwise. I had estimated that Amchitka supported somewhere between about 6,500 and 8,000 or more sea otters in the 1970s, using the following population-assessment technique. I first selected four stretches of coastline around the island, each about 10 miles in length, within which I repeatedly counted the animals from shore on days when viewing conditions were excellent. I then conducted full-island aerial surveys from a helicopter, noting the proportions of the total counts that occurred in the shore-based areas, and adjusted the total counts upward according to the ratio of shore counts to aerial counts (Estes, 1977).

I didn't have access to a helicopter in the early 1990s, but in 1993 I had a large field crew at Amchitka and I had them survey the entire island from the shore. I divided the 12-person field crew into six teams of observers, making sure that each team contained at least one experienced otter-counter. I then divided the coast of Amchitka Island into six segments, each about 20 miles in length. During a lucky stretch of clear, calm weather, we were thus able to survey the entire island in three days.

We counted just under 4,000 independent sea otters. The number seemed low, but wildlife population surveys are notoriously inaccurate and I had only a single data point for the early 1990s. So I wondered—had the population declined by half, were otters somehow missed during the whole-island shore survey, or had I overestimated the population in the early 1970s? I left Amchitka in 1993 feeling deeply unresolved. I knew that something had changed, but I didn't know what.

I would have left the Aleutians in this state of mind were it not for Dan Boone, an employee of the U.S. Fish and Wildlife Service (USFWS) who managed the Aleutian Islands Unit of the Alaska Maritime National Wildlife Refuge. The U.S. Navy, which occupied Adak, and the U.S. Air Force, which occupied Shemya, both had received funding through the Department of Defense Legacy Resource Management Program for environmental research on their respective military bases. The refuge wanted some of this work to be done on sea otters, and Dan asked me to lead that effort. Frankly, I was luke-warm to his proposal and probably would have declined had it not been for the peculiar results from our recent project at Amchitka. So I agreed, intending to expand my studies of sea otter population and behavior to Adak. I wanted to know whether the strange things we had seen at Amchitka were also occurring at Adak. And if so, I hoped to figure out why the otters' behavior and life history had changed so much since the early 1970s.

I began at Adak in 1994 as I had at Amchitka in 1991—by capturing and tagging sea otters. Field support from the navy and the USFWS made the work at Adak easier than it had been at Amchitka. In the mid-1980s, a team of four of us had captured 50 otters at Adak in less than three days. I thus expected the capture and tagging part of the study to go smoothly and quickly. But as it had been at Amchitka, the otters were not so easy to capture. Instead of making all our captures within a week, we ground away day after day for nearly two weeks to obtain just 37 otters. This was many fewer than the 60 I had planned for. I kept hoping our fortunes would turn for the better, but instead they turned for the worse.

I had retained Carolyn McCormick, a veterinarian from Anchorage, to drug and attach transmitters to the captured sea otters. Carolyn had worked with me at Amchitka in 1991. She was good with the animals and good in the field. I had retained a second veterinarian (Vicki Vanek from Kodiak) on the Amchitka project because of the heavy workload and to serve as a backup if anything happened to prevent Carolyn from finishing the capture-and-tagging phase of the study. We would be capturing fewer animals at Adak, and I didn't have funds to pay two veterinarians. So I decided to work with a single verterinarian this time, recognizing that we depended entirely on Carolyn but also thinking the chances of that becoming a problem were remote. And it wouldn't have been a problem if the captures had proceeded as quickly as I thought they would.

Early one morning midway through the second week of captures, Carolyn knocked on my door with bad news—her fiancé had been seriously injured in a farming accident. She was distraught and told me she would have to leave for home as soon as possible. While I understood her feelings and concurred with her decision, I also realized that her departure would mark the end of our capture efforts on Adak for that year. We would simply have to make do with the 37 animals we had thus far captured and marked for study.

But the next flight out of Adak wouldn't be for another four days. Carolyn took it in stride, offering to help capture and tag as many more otters as possible in the interim. However, as dawn broke that day, the wind began to freshen with an approaching storm, which meant that we probably wouldn't be able to capture any more otters from the exposed outer-coast habitats where I wanted to study them. There were just two options—call it quits or move our capture operation into nearby Clam Lagoon, which offered protection from the weather and building seas. About 100 sea otters lived in Clam Lagoon, but I had avoided working there because the habitat and prey base differed from the outer coastal areas that characterized most of the Aleutian Islands. In fact, Clam Lagoon was a geological anomaly, created during the Pleistocene by a terminal glacial moraine that was subsequently breached by erosion and the forces of the open sea to form a sort of marine lake, roughly two miles in length by a mile wide and connected to the open sea by a shallow tidal channel. Although the lagoon was thoroughly flushed by the waxing and waning tides, its ocean entrance could be waded across during low *spring tides* and was barely navigable by small skiff at high tide. Clam Lagoon proper was easily navigable and provided a lee to wind and rough seas from any direction.

High tide would occur later that morning, so I quickly decided to move our capture operation to Clam Lagoon. It was that or nothing, and maybe we would learn something from the otters there. I knew these animals would be feeding on a different prey base from the one otters exploited on the outer coast, but I hoped their behaviors and life histories would otherwise be similar enough to be useful, elevating our sample size if nothing else.

We moved our skiff and capture gear into Clam Lagoon; by early afternoon we had deployed the capture nets, and by evening we had already captured six otters. In less than three days we had captured and tagged 11 more otters. This high capture rate was much as it had been at Amchitka

FIGURE 10.1. Tim Tinker (right) with the author, surveying sea otters at Attu Island. (Photo by Norman S. Smith)

and elsewhere on Adak in earlier years. At the time I didn't think much of this, but later it helped me understand what was happening in the open-sea habitats outside Clam Lagoon.

After Carolyn's departure, we began the long process of studying the marked otters. I too would be leaving Adak in a few more weeks, after which the monitoring would be done by Tim Tinker and Julie Stewart. I had met Tim (figure 10.1) several years earlier at a Marine Mammal conference in Galveston, Texas. At the time I was looking for someone to serve as a field technician at Amchitka and had posted an advertisement on the conference bulletin board, to which Tim responded. He had completed a master's degree with Kit Kovaks at the University of Waterloo and had field research experience with harp seals in Nova Scotia, as a fisheries observer in the Bay of Fundy, and as a tree planter in British Columbia. He was keen for the job at Amchitka and we quickly sealed the deal. As the Amchitka project was winding down and the Adak project was ramping up, Tim expressed interest in staying on. Julie was also an experienced fieldworker, so Tim and Julie were the natural choice to work on the Adak project.

Before leaving Adak later that summer, Tim and I surveyed the sea otter population by running a skiff around the island's perimeter, noting the location, group size, and reproductive status (i.e., presence or absence of a dependent pup) of every animal we saw. I had begun using this survey method a few years earlier because I no longer had access to a helicopter and it seemed to produce consistent numbers. The goal was not to obtain a population estimate but rather to document spatial patterns and time trends in the distribution and abundance of sea otters. This was our first complete survey of Adak. The numbers were lower than I thought they should be and, for a section of Adak, lower than those we had obtained in a partial survey in 1991. The possibility that sea otter numbers at Adak and Amchitka were in decline was beginning to dawn on me, but the data weren't conclusive enough to infer anything of substance.

Tim and Julie were given a protocol for field measurements, which included weekly relocations of the tagged sea otters, systematic measurements of their diet and activity, beach surveys for carcasses, and occasional skiff surveys of the abundance and distribution of otters within an intensive study area, which included Kuluk Bay and Clam Lagoon. Tim wrote or called periodically with updates and to discuss the findings. During one of these calls, he noted that two of the radio-tagged otters in Kuluk Bay had gone missing after a group of killer whales had passed through the bay. And then, recalling when several other tagged otters had gone missing, he realized that those, too, had followed sightings of killer whales (figure 10.2). Although the possibility that sea otter populations were being influenced by killer whale predation began to dawn on us, I was skeptical. Our evidence of an otter population decline was preliminary and still fairly thin, and we had little direct evidence of killer whales preying on otters.

In fact, prior to 1991, there had been no confirmed attacks by killer whales on sea otters. I had spent thousands of hours in the field with these animals and had never seen an attack. Admittedly, though, I also rarely saw killer whales in coastal waters of the Aleutian Islands during the 1970s and '80s. They may have been around, but if so they were either exceedingly cryptic or spent most of their time farther from shore. But eventually, in the summer of 1992, while collecting foraging data near St. Makarias Point on the south coast of Amchitka, Brian Hatfield, a member of my field crew, saw a killer whale rush into a kelp bed in apparent pursuit of an otter. The otter took what Brian interpreted as evasive action, and then it was gone. He never saw the otter in the jaws of the killer whale and was thus unsure if it had been swallowed underwater or escaped unseen.

The next day Brian observed a similar attack in the same area, although this time he also saw what looked like body parts of the otter floating on the

FIGURE 10.2. Killer whale in kelp forest at Kiska Island. (Photo by Norman S. Smith)

ocean's surface after the attack. A bald eagle had swooped down on the remains, carrying what appeared to be a section of intestine back to its nest. When Brian told me about this observation, I decided to climb into the eagle's nest in search of the remains. But the nest contained several large chicks, and whatever he saw the eagle carrying had already been consumed. I was convinced that Brian had witnessed attacks by killer whales on sea otters, but I considered this an unusual and ecologically unimportant event.

But sightings of these attacks on sea otters continued to mount. On Adak, a killer whale opened its jaws and engulfed an otter right before Tim's and Julie's eyes. While gathering foraging observations near Lucky Point on Adak, several of our volunteer field crew witnessed another attack that involved a more purposeful and coordinated series of behaviors by several killer whales. In this instance, a small group of otters was hauled out on an offshore shoal that had become exposed at low tide. The weather was calm and the observers saw the killer whales coming from a distance. The latter stopped several hundred meters short of the shoal, at which point two of them sank beneath the surface and reappeared on the other side of the shoal.

The lone remaining one then rushed the shoal at high speed, turning abruptly just before reaching it and thereby creating a wave that washed the otters off the rocks and into the jaws of the two killer whales waiting on the other side. It must have been quite a spectacle because the observers, as the story goes, left their inflatable skiff on the beach and *walked* back to camp.

Tim and I surveyed Adak again in the summer of 1995 and counted many fewer otters than in 1994. Together with results from the partial survey in 1991, these data strongly indicated a substantial population decline. I wasn't absolutely sure because we had only one partial and two complete surveys of Adak.

The data Tim and Julie had acquired during the preceding year were also intriguing, especially the contrast between otters in Kuluk Bay and Clam Lagoon. About 10 percent of the tagged animals in Clam Lagoon had died or disappeared, whereas more than 60 percent of the marked animals in Kuluk Bay had gone missing over the same period. Our tagged animals rarely moved between Kuluk Bay and Clam Lagoon, and we were certain, given the animals' home-range sizes and searches of Adak's coastal perimeter, that the missing animals from Kuluk Bay had not moved elsewhere.

Unfortunately, the Legacy funding was for a single year, and although I had been told it would likely be extended for at least a second year, it was not. So we packed up and left Adak, planning to use what funds remained to conduct another survey the next year. When we returned to Adak in 1996 to survey the population and locate the remaining radio-tagged animals, we found that the population had continued its sharp decline, and the loss rate of tagged animals in Kuluk Bay continued to be much higher than that in Clam Lagoon. By this point I knew that the Adak sea otter population was in steep decline, and I was fairly sure the explanation somehow lay in the contrast between Clam Lagoon and Kuluk Bay. I suspected killer whales but didn't have enough supporting evidence to embrace that view with confidence. And besides, another potential explanation for the decline had surfaced: environmental contaminants.

My interest in contaminants grew out of discussions with Wally Jarman, a toxicologist at UC Santa Cruz, and Corinne Bacon, a recent graduate who had worked for Wally and wanted to continue with him for a master's degree. Corinne was interested in sea otters, Wally's lab was set up to conduct contaminant analyses, and I had access to sea otter tissue samples from various places across the Pacific Rim. Corinne's thesis research would compare tissue concentrations of *PCBs*, DDT, and several other persistent organic compounds among three sites: central California, southeast Alaska, and the western Aleutians. Our initial interest was primarily in California, and specifically whether contaminants might explain why the California sea otter population was failing to recover. Corinne's contaminant survey would not answer that question, but it would tell us if contaminants were a viable hypothesis. We chose the Aleutians as a point of comparison mainly because we expected such a remote and purportedly pristine area to be about as free from anthropogenic contaminants as any place within the sea otter's geographic range. Furthermore, we knew that sea otter numbers were increasing rapidly in southeast Alaska and thought that the populations were stable at high densities in the islands (Amchitka and Adak) from which we had carcass samples in the Aleutians. Whatever kinds and concentrations of contaminants we would discover in southeast Alaska and the Aleutians, sea otters there ought to be good for establishing minimum "safe" thresholds for the species.

We were shocked by the results. As expected, contaminant concentrations were relatively low in southeast Alaska and significantly higher in central California. But PCB concentrations in Aleutian otters were almost 40 times higher than they were in southeast Alaska and more than two times higher than in California (Bacon et al., 1999). The Aleutians were not the pristine and chemically clean environment we had imagined them to be. We didn't know what this meant for the otters. But if PCBs were in any way problematic for sea otters in California, it seemed they would be even more so in the Aleutians.

These data and their implications emerged with the completion of Corinne's laboratory work in 1996. Sea otter numbers at Adak continued their steep decline. Although I was having a difficult time reconciling how contaminants could have had such a sudden and profound influence on sea otters, there was no denying the need to further explore that possibility. The navy was especially concerned and asked us to do just that.

The obvious next step was to chronicle contaminant concentrations across the Aleutians and southeast Alaska in greater detail. We decided to do this by measuring contaminant loads in the blood of living otters and in the tissues of other common species at a variety of locations around Adak and more broadly across the Aleutians. These measurements would provide a comprehensive view of the spatial extent of the sea otter's decline and any co-occurring patterns of contaminant loads, in turn telling us whether the otter's decline was restricted to Adak or more widespread across the Aleutians, and whether elevated contaminant levels corresponded with whatever the pattern of decline turned out to be. The navy was particularly interested in knowing whether they were to blame for any contaminant-related problem. Our proposed sampling program would answer that question.

We began these additional studies of contaminants in the summer of 1997. Although several more years would be needed to complete the sampling and analyze and interpret the data, the critical questions had been answered by the following winter. Resurveys of sea otters at several islands from which I had longer-term records established that populations on these islands had also declined (Doroff et al., 2003). The contaminant measurements from sea otter blood and intertidal mussels further established that elevated PCB concentrations were localized around the sites from which the carcasses in Corinne's study had been collected (Kuluk Bay on Adak and Constantine Harbor on Amchitka). Levels elsewhere across the Aleutians were low (Jessup et al., 2010). These data convinced me that the sea otter's decline, which by now I realized was widespread across at least the central and western Aleutians, was not being caused by environmental contaminants.

Nutritional limitation—reduction in prey abundance or quality—was another potential explanation for the decline. The 1987 reef surveys (see chapter 7) provided information on sea urchin abundance and size at Adak and Amchitka from a period just before the sea otter's decline began. I decided to resurvey the sites at Adak in 1997. By this point the otter population had declined by nearly 90 percent, and I thought that the reef resurvey data might help me better understand the cause of the decline. If nutritional limitation were the cause, I expected to see, among other things, a decline in sea otter prey. Conversely, inasmuch as sea otters also limit their prey abundance, I expected to see an increase in prey abundance if the decline was not being caused by nutritional limitation. I didn't expect to see a change in kelp distribution or abundance, mainly because I hadn't seen any hint of a kelp forest phase shift through 1994, the last year I had dived extensively at Adak. Persistence of the urchin phase state around Attu through more than 20 years of sea otter population increase had also left me with the notion that phase shifts had long lag times in response to changes in sea otter abundance.

As we began the reef resurvey of Adak in August 1997, I was unprepared for what I was about to see. Urchin biomass density had increased by nearly tenfold since 1987. Most of the sites had shifted to the urchin-dominated phase state, and kelp populations were in the process of being devoured by urchins at the few that hadn't. I drew two conclusions from these observations and data. One was that the otter's decline at Adak was not a consequence of nutritional limitation. And second, the decline of otters at Adak had resulted in a kelp forest phase shift. We resurveyed Amchitka Island in 1999, and there, too, the reefs had shifted from kelp forests to urchin barrens. Although the cause or causes of the sea otter's decline remained uncertain, the decline's consequences were clear.

In the fall of 1997, Tim Tinker and I began a process of assessing what we could infer from available information about the cause of the decline. We did this by first assembling two lists: one contained every possible cause that we could imagine, and the other was a detailed summary of everything we knew about sea otters and their coastal ecosystems in the Aleutians. We then combined the two lists in a matrix, with the potential causes in one dimension and the information on otters and their associated ecosystem in the other. The idea was to determine whether one of the hypothesized reasons for the decline would emerge as most likely from the weight of evidence. In particular, we explored this matrix for what we thought of as "fatal flaws"—data or observations that were inconsistent with a particular causal explanation. In these cases, we asked ourselves two further questions: Do we believe that the data or observations are robust, and are we sure about the logic underlying our conclusions of fatal flaws when the data set and any particular hypothesis have failed to align? Everything was on the table at the onset of this exercise. But once we identified a fatal flaw, that particular explanation was taken off the table. This exercise was helped immeasurably by the data from Clam Lagoon, within which the otters had not declined but outside of which they had. Any explanation for the decline would have to account for that pattern.

We knew from the outset that the decline was a result of elevated mortality—not reduced reproduction, and almost certainly not emigration. We were able to immediately exclude food and contaminants, as explained above. Conceivably, we may have looked at the wrong contaminants, or the decline may have been caused by some naturally occurring *biotoxin*. But any such explanation was inconsistent with the wide-ranging decline on one hand and the persistence of high otter densities in Clam Lagoon on the other. The continuous, tidally driven exchange of sea water between Clam Lagoon and the adjacent open ocean would have exposed the otters to waterborne toxins or contaminants. A similar argument and conclusions applied to the hypothesis of disease. We had looked at the living animals for health-related problems, in particular exposure to disease. The animals all appeared to be in excellent health (Hanni et al., 2003). There was no evidence of exposure to any known disease, and the absence of beach-cast carcasses argued further against disease or toxins.

This exercise led to the exclusion of all but one reasonable explanation for the sea otter's decline, and that was predation. We suspected killer whales because of both the striking increase in sightings (from an occasional observation every few years to what were often multiple sightings per day) and the observed attacks on sea otters during the period of the otter's decline. But the number of observed attacks (six) seemed far too few to account for the large number of deaths that were needed to drive the decline. It was also possible that some other predator or predators, such as sharks or humans, were contributing to or causing the decline. We were almost certain the decline was not caused by direct human exploitation or incidental loss from *bycatch* in some fishery. Surely we would have seen or heard of these things if they were important, but we had not. Sharks were another possibility, but to date we have discounted this because of the absence of evidence that either sleeper sharks or salmon sharks (the two most common large shark species in the Aleutians) eat sea otters, that their abundance has changed, or that they were even present in sufficient numbers across the Aleutians to be able to cause such a rapid change in sea otter abundance.

Although the killer whale hypothesis was intriguing, I was troubled by two outstanding questions. The first of these was how to reconcile the small number of observed attacks with the large numbers of deaths that would have been required to drive sea otter populations downward at the observed rate of almost 25 percent per year. And the second was whether there were enough killer whales in the area to have eaten so many otters. To answer these questions, I sought assistance from two colleagues at UC Santa Cruz, population ecologist Dan Doak (figure 10.3) and comparative physiologist Terrie Williams (figure 10.4).

Dan helped us estimate the number of "additional" otter deaths needed to account for the observed declines. We first had to specify a geographic area for analysis. The precise borders weren't critical, so long as (a) the number of sea otters within the area before the decline could be reasonably estimated, (b) the rate of decline was similar throughout the area, and (c) our field observations were a representative sample of the area. We ended up using the area from Kiska Island in the west to Seguam Island in the east because we knew that otter populations had been at or near carrying capacity across this region at the decline's onset, and so we could estimate the predecline population size

FIGURE 10.3. Daniel F. Doak, whose modeling work has been essential to my understanding of predator–prey dynamics.

within the region. Sea otter populations had evidently declined throughout the region, and our field observations were presumably representative of the larger region.

From years of studying sea otters, I knew their age-specific fertility rates. Age-specific mortality rates were less certain, but we were able to combine the data we did have with a simulation modeling approach to come up with an age-specific mortality schedule that, when combined with the age-specific fertility rates, would result in a stationary population (i.e., one for which *lambda* [λ] = 1). By making the simplifying but seemingly reasonable additional assumption that the increased probability of death from killer whale predation was spread equally across all ages, we could calculate the overall number of additional deaths that would be needed to drive the sea otter populations between Kiska and Seguam downward at the observed rate. That number turned out to be just over 40,000 sea otters. I then asked how many of these deaths we would have expected to see if they were all caused by killer whale predation, given the number of person-hours that had been spent in the field and the area within which an observer, perched on a cliff overlooking the ocean, might expect to see and recognize an attack that occurred within that area while he or she was watching. Based on the estimated

FIGURE 10.4. Terrie M. Williams, whose work on metabolism and energetics helped us understand the energy and material values of marine mammals as prey for killer whales.

distances between the field observers and the few observed attacks, I figured this distance to be about 1 kilometer.

This analysis contained many uncertainties and simplifying assumptions. For example, the distance at which a shore-based observer might be expected to see an attack would necessarily vary with weather and viewing conditions. But unless we had made a gross error or miscalculation, none of that mattered for my purpose, which was simply to determine whether the number of attacks we saw was in the ballpark of the number we would expect to see. Given that perspective, the answer was highly informative. If the 40,000 additional deaths occurred more or less randomly across space and time, I calculated that we would have expected to observe five of them. We had seen six attacks. These are small numbers, and their uncanny similarity may be little more than coincidence. But clearly, the small number of observed attacks was at least not inconsistent with the assumption that all the sea otter deaths in question had been caused by killer whale predation.

We were satisfied with the estimates and analysis just described, but the fact remained that more than 40,000 sea otters had disappeared over the course of a few years. Were there enough killer whales in the central and western Aleutian Islands to eat that many otters? Our uncertainty was compounded by the knowledge that most killer whales don't eat marine mammals; they mainly consume fish. Years of field study by killer whale biologists had revealed at least two distinct killer whale ecotypes—the "residents" (or fish eaters) and the "transients" (or marine mammal eaters)—and residents were typically about 10 times more common than transients. If the same were true in the Aleutian Islands, we thought there might be too few transient killer whales to eat so many otters. Our concern was compounded by the uncertain number of transient killer whales in the central and western Aleutians. We knew they occurred in that region, but we didn't know how many. Rather than try to figure that out, we simply asked how many transient killer whales would be needed to eat so many otters.

Terrie approached this question in the following way. First, she asked how much energy is needed to fuel a killer whale. By combining the broadly predictive allometric relationship between body mass and *field metabolic rate* in mammals with the few scant data from studies of captive killer whales, she was able to come up with a range of values that had to be within severalfold of the correct answer. This told us about how much energy a wild killer whale needs to fuel its existence.

The next step was to estimate the energetic value of an otter. To do that, Terrie obtained an otter carcass and sent it to the UC Davis School of Veterinary Medicine to be homogenized in their "large animal blender," a ghoulish undertaking that produced a homogenate, which Terrie then ran through a bomb calorimeter to determine the energy density of an entire sea otter. As it turned out, the caloric density of sea otters was about mid-range among the various killer whale prey species, which dispelled the common misperception that killer whales would obtain little of value from eating sea otters. Armed with this number, the average body mass of a sea otter, and the *assimilation efficiency* of a killer whale (estimated at about 80 percent, a standard and fairly consistent value for mammalian carnivores), Terrie could then estimate the consumption rate of otters by killer whales.

Here again, we were astonished by the results. If we assumed that killer whales fed on nothing but sea otters, just three adult killer whales could have accounted for all the deaths indicated by our demographic analysis. Even if we relaxed the dietary assumption to a hypothetical situation in which the sea otters were just 1 percent of the killer whales' diet, 300 transient killer whales would account for all the losses. We knew there were many more than three transient killer whales in the Aleutian Islands, and we were fairly sure there were more than 300 of them. These analyses convinced us that killer whales could easily have eaten all the missing otters.

I wasn't entirely convinced that the otter's decline had been caused by killer whale predation, but I was leaning more strongly in that direction as the evidence mounted. I was virtually certain the declines had little or nothing to do with toxins, diseases, or food limitation. Food limitation should have been accompanied by reduced food abundance, poor body condition, and more effort invested in foraging, but our data showed just the opposite. Starvation, disease, or pollutants should have produced large numbers of stranded carcasses, but there were none; nor was there evidence of toxins or disease agents in the otters themselves. Beyond that, it was inconceivable to me how disease or pollutants could have reduced populations so uniformly over thousands of miles of shoreline across the central and western Aleutians while leaving the small enclave of otters in Clam Lagoon unaffected. I couldn't believe the declines had been caused purposefully or inadvertently by people, because there were so few people in the area and I had no reason to suspect that the numbers or behavior of these people had changed.

The evidence, such as it was, all pointed to killer whales. The presence of these animals in coastal waters of the Aleutian archipelago had increased substantially at about the onset of the decline; we had seen killer whales attacking sea otters during the decline, something that had not been seen before the decline; killer whales were unable to enter Clam Lagoon, which was consistent with the unique lack of decline in this small area; killer whale predation would explain the absence of stranded sea otter carcasses; and our demographic and energetic modeling analyses were consistent with killer whale predation.

There was a problem, however, and that was my own sense of scientific argument. I had been trained throughout my professional life to be skeptical and self-critical. So while it was easy enough for me to point out the patterns, I was uncomfortable making the argument that killer whales had caused the sea otter's decline, even in a cautious and appropriately nuanced manner. I thought that regardless of how I presented the data and told the story, it would be controversial.

That feeling began to change as I spoke of this work at meetings and seminars. People I respected urged me to publish the story in a visible outlet, arguing that our findings were newsworthy and that my only responsibility was to present the data and my interpretation of them in a fair and truthful manner. Doing that would allow others to be the judge. And if further study or analysis showed the story to be wrong, so be it.

I submitted a report of these various findings to the prestigious journal *Science* in early 1998. The manuscript was promptly declined, without review. A few months later I happened to mention this to Bruce Lyon, a colleague and fellow faculty member at UC Santa Cruz. Bruce was astonished that the manuscript had not at least been sent out for review, and he urged me to challenge the editors' decision. So I resubmitted a slightly revised version with the more provocative title "Killer whale predation on sea otters linking coastal with oceanic ecosystems." The editors at *Science* agreed to reconsider the manuscript and sent it out for review; the reviews were strongly positive and the paper was published in early October of that year (Estes et al., 1998).

I expected that the media would take note of our findings, but I was unprepared for the magnitude of their interest. For weeks it seemed that I did little but speak with journalists. The story was everywhere—in newspapers and magazines, on the radio, and eventually in various textbooks. And while I'm sure there was grousing, the blowback I had feared from a hypercritical scientific community didn't happen. But the publication of this paper started me down a new path. My life as a scientist would forever be changed.

ELEVEN

Megafaunal Collapse

THE COLLAPSE OF SEA OTTER POPULATIONS and their coastal ecosystems in southwest Alaska inevitably led me to wonder why, but the search for answers was challenging. The utterly surprising nature of the collapse—when and where it happened—made me realize that the ecosystem I had been studying for more than a quarter of a century was organized around processes I didn't understand. But Pandora's Box was open and I had to press on.

Where to start? The path forward, at least in my mind, went through killer whales (see chapter 10). But I also worried that if I made killer whales the foundation, I might end up building a house of cards. There had been no formal challenges to the killer whale hypothesis, but it was still just that— a working hypothesis. If it proved correct, then any further efforts to understand killer whales might lead me down the correct path of understanding. But if it proved wrong, then any such effort would be misdirected. I worried about this. I didn't really doubt the killer whale hypothesis by this point, but for the time being I had no way to assess its likelihood of being correct, except by way of my own internal logic and the fatal inconsistencies of every alternative hypothesis. Never underestimate the worry and sense of responsibility that comes with guiding a large research project and leading one's coworkers on a possibly fruitless quest. But rather than succumb to paralysis from the fear of being wrong, pressing on in an effort to understand why killer whales suddenly gobbled up the sea otters seemed like the right thing to do.

The first step in this endeavor was to figure out why the sea otter–coastal ecosystem collapse occurred when and where it did. I knew that the sea otter's decline had begun in about 1990, give or take a few years. Why hadn't it begun a decade earlier or later? I knew that the decline could have been caused by the smallest variation in transient killer whales' foraging behavior—a change of a percentage point or two in the composition of their diet, or a shift by just several individuals to eating sea otters. Thus, the timing of the decline could have been set by nothing more than some transient killer whale's fortuitous discovery of sea otters as an abundant and easy mark. From such a discovery, the behavior might easily have spread to other members of the family unit (*matriline*), which in turn is all that would have been needed to cause the decline. I thought of this as the "fortuitous discovery hypothesis." Such an event could have occurred in an otherwise static world. The diet and foraging behavior of transient killer whales often differ among matrilines, and novel behavioral discoveries that led to significant ecological change and individual variation are known from numerous consumer species. And we had seen the same individual killer whales prowling the coastal waters of the central and western Aleutians year after year. But beyond those few tidbits of knowledge, the fortuitous discovery hypothesis implied that the sea otter's decline occurred in an environment that was otherwise ecologically static. That didn't mean the hypothesis was wrong, but I had no strong basis for arguing that it was right. I couldn't think of anything further to do to put the fortuitous discovery hypothesis to additional tests.

I might have stopped there had I not already known that the marine ecosystem of southwestern Alaska was *not* static, and indeed it had changed in a manner that could have caused killer whales to prey more intensively on sea otters. The environmental change that captured my attention was the previous collapse of Steller sea lion populations. While the spatial extent of the sea lion's decline was roughly similar to that of the sea otters, it had occurred about a decade earlier, beginning sometime in the late 1970s or early '80s and bottoming out in the late 1980s or early '90s. Transient killer whales were known to prey on sea lions. I thus imagined that some of the sea lion–eating killer whales had turned to eating sea otters as sea lions dwindled.

I thought of this as the "sea lion switching hypothesis." In contrast to the fortuitous discovery hypothesis, this hypothesis was more compelling because it fit both the expectations of *foraging theory* and events that had actually occurred and been seen in nature: killer whales eating sea lions; sea lion abundance declining; killer whales beginning to eat sea otters; and sea otter populations subsequently collapsing. The declines of the sea lion and sea otter may have been unrelated, but that seemed unlikely. I thought it more likely that these events were linked by killer whale predation. Now, almost 20 years later, I still think that's the most reasonable explanation for what happened.

If this line of reasoning were correct, then understanding the collapse of sea otter populations was ultimately linked to understanding the sea lion's decline. I hadn't studied sea lions, but the consensus among those who had was that the declines were, in one way or another, the result of nutritional limitation. This supposition had apparently arisen, in significant part, from a widely held belief in oceanography, fisheries biology, and marine mammalogy that dynamic processes in the open sea are commonly characterized by bottom-up control (see chapter 2). At the time, the question was not whether nutritional limitation had caused sea lions to decline but how that came to be. Two camps had developed around the issue, one of which thought that the purported nutritional limitation in sea lions was caused by competition with fisheries while the other favored a late-1970s shift in the *Pacific Decadal Oscillation* (PDO).

Although I hadn't studied the evidence carefully, large amounts of money, time, and effort had been spent in trying to understand the sea lion's decline, and I assumed that the people who had been engaged in this effort were correct in attributing the collapse of sea lion populations to bottom-up forcing. Our 1998 report to *Science* on the collapse of sea otter populations (Estes et al., 1998) embraced that view by proposing the following sequence of events. Sea lion populations declined during the 1980s because of nutritional limitation, which may have been caused by fisheries, the PDO, or the combined effects of both. As sea lion numbers dwindled, the killer whales that fed on sea lions switched or expanded their diets to include sea otters, thereby driving sea otter numbers sharply downward, which in turn caused the coastal ecosystem to shift from a kelp- to an urchin-dominated state. It was a slick explanation for all that had happened. But I would soon come to believe that an important part of the story was wrong.

The difficulty was in a lack of convincing evidence for nutritional limitation in sea lions. For one thing, there was no evidence that the sea lion's food resources had dwindled. If anything, overall finfish biomass in the Bering Sea and North Pacific Ocean had increased as the sea lions declined. Nonetheless, some sea lion researchers continued to invoke nutritional limitation by claiming or imagining a decline in prey quality, the idea being that *gadids* (especially walleye pollock), which composed a large proportion of the regional fish biomass, were too rich in protein and too poor in lipid to support sea lions. This idea became known as the "junk food hypothesis." Proponents of the junk food hypothesis, using large-scale dietary patterns as supporting evidence, had concluded that the magnitude of local declines in sea lion populations was strongly correlated with dietary diversity (the most depleted populations had the least diverse diets) and consumption of pollock (the most depleted sea lion populations ate proportionally more pollock).

I was never compelled by this line of reasoning and began to suspect that proponents of the junk food hypothesis had turned their thinking and arguments to support a preconceived view of nutritional limitation, while at the same time reconciling how this might happen when there were lots of fish in the surrounding oceans. My initial doubt grew out of an alternative explanation for the clear relationship between dietary diversity and the depth of decline in local sea lion populations, one that followed from the most basic tenets of *optimal foraging theory* and its associated predictions of the relations between prey choice, consumer population status, and prey abundance. In short, optimal foraging theory predicts that dietary diversity should increase with increasing consumer numbers and a resulting depletion of the consumer's preferred prey. If sea lions were capable of limiting their prey and if sea lion numbers were being driven downward by something other than nutritional limitation, one would expect prey abundance to increase and dietary diversity to decline. This pattern had been seen repeatedly in other consumer–prey systems; for sea lions it only requires that pollock are not junk food.

Others were beginning to question the nutritional limitation hypothesis for other reasons. If nutritional limitation were responsible for the sea lion's decline, one would expect to find some evidence of starvation. But nutritionally stressed sea lions in the areas of decline had not been found, and those animals examined on haul-outs and rookeries appeared to be well fed. Studies of diving and foraging behavior showed that sea lions within the region of decline spent less time feeding and dove to shallower depths than animals from populations in southeast Alaska that had not declined. Captive feeding studies of both sea lions and harbor seals failed to support the junk food hypothesis, in that animals raised on pollock diets grew and gained weight as rapidly as animals raised on other diets.

These conflicting results between the nutritional limitation hypothesis and information from nature were beginning to resonate in a confusing way with many scientists, managers, and policy makers. The proponents of nutritional limitation stuck to their guns, arguing that the science was still equivocal and that a precautionary approach that limited fishery take was needed to protect and recover the sea lions. But the Bering Sea–North Pacific ground fishery was the single most economically important fishery in the United

States, generating revenues in excess of several billion dollars a year. Politicians and fishery managers were loath to limit the fishery on behalf of sea lion welfare in the absence of better evidence that the fishery was negatively affecting these animals.

In an effort to resolve these conflicting perspectives on underlying reasons for the sea lion's decline, the National Research Council (NRC) was asked to undertake an independent study. The NRC committee comprised a diverse array of scientists—marine ecologists, fishery biologists, demographers, physiologists, oceanographers, climate scientists, veterinarians, and the like—none of whom worked for NOAA Fisheries (the office of the National Oceanic and Atmospheric Administration responsible for sea lion management) or had a vested interest in the questions at hand. I served on that committee. We read the published literature and the unpublished technical reports; we heard testimony from scientists who had done the research on sea lions; and we examined as much of the raw and digested data as the owners of those data were willing to give us. In the end we concluded that the sea lion's failure to recover was almost certainly not the consequence of nutritional limitation.

NRC study reports are often conservative in their assessments of controversial issues and conflicting points of view. For that reason and because of the near absence of research during the period of decline, we softened our written assessment of the cause of the sea lion's decline with caveats that acknowledged uncertainty. But to my mind, there were simply too many inconsistencies between the data and the arguments to support the nutritional limitation hypothesis. Food abundance or quality may have been involved somehow, but it seemed to me that those who embraced that view did so more out of principle than because of the evidence. I had become more skeptical than ever that the decline had anything to do with nutritional limitation in sea lions.

If nutritional limitation hadn't caused the sea lions to decline, what had? Having recently concluded that the sea otter's decline was caused by killer whale predation, I couldn't help but wonder whether killer whales had eaten the missing sea lions as well. Although this hypothesis had been largely dismissed by those who had studied sea lions and killer whales, similarities between the decline of the sea otter and that of the sea lion intrigued me. Transient killer whales often preyed on sea lions, so the idea that the sea lion's decline had been driven, to some degree, by killer whale predation was at least plausible. The area over which the sea lions had declined—from the central Gulf of Alaska westward through the Aleutian archipelago—was broadly similar to the area over which the sea otters had declined. Even though the sea lion's decline had occurred a decade earlier, the trajectories and magnitudes of its decline and that of the sea otter were similar. And as with the sea otter, sea lion carcasses or remains were seldom found.

Unfortunately, no one had been studying sea lions or killer whales in southwest Alaska during the period of major decline in the 1980s; those studies didn't begin until after the decline had largely run its course. But energetic and demographic modeling could be used to ask the same question about sea lions that we had asked about sea otters. That is, was predation a feasible explanation and, if so, how many killer whales were needed to eat enough sea lions to drive the decline?

Dan Doak and Terrie Williams conducted these analyses for killer whales and sea lions, just as they had for killer whales and sea otters. This time we weren't so surprised. Their analyses indicated that in an ocean environment in which all else was constant, the sea lion's decline across all of southwest Alaska could have been caused by fewer than 40 killer whales switching entirely from not eating sea lions to a diet composed entirely of sea lions, or by a roughly 10 percent increase in sea lions in the diet of the entire population of transient killer whales (Williams et al., 2004).

I wasn't sure then and I'm not sure now that the sea lion's decline was caused either in whole or in significant part by killer whale predation. But in contrast to nutritional limitation, there were no fatal inconsistencies between the predation hypothesis and available data. I was less certain for sea lions than for sea otters that the population decline had been caused by killer whale predation. However, this uncertainty was due to a lack of information rather than conflicting evidence. The evidence was and still is more consistent with predation than with any other potential explanation of the sea lion's decline. And as with the sea otters, all other reasonable hypotheses had one or more fatal flaws when lined up next to the available data. I still see it that way and will stand by my view until someone provides evidence or a compelling argument to the contrary.

Even though the purported underlying cause of nutritional limitation in sea lions was debated, the nutritional limitation hypothesis was satisfying to its proponents because it provided a logical starting point for the sea otter's decline. That starting point, in their view, was either overfishing, physical oceanographic changes associated with the PDO state shift, or some combined or interactive effect of the two. Once I had rejected the nutritional

FIGURE 11.1. Alan M. Springer in the Pribilof Islands.

limitation hypothesis, tough questions about the ultimate causes of the sea lion's and sea otter's declines were again on the table.

I was essentially back to square one, asking the same question of killer whales and sea lions that I had asked earlier of killer whales and sea otters. But while the decline of the sea lion provided a reasonable explanation for why the sea otter's decline began when it did, a similar explanation was missing for the onset of the sea lion's decline. If it was caused by killer whale predation, why did this happen when and where it did? Without an answer to that question, the predation hypothesis was as hollow as any other.

A key insight into the sea lion's decline came from Alan Springer, a broad-thinking marine scientist from the University of Alaska with decades of experience in the North Pacific, Bering Sea, and western Arctic regions (figure 11.1). I had known of Alan and his work for years, but we met for the first time at a NOAA-sponsored workshop on Steller sea lions in 1999. About a year later, Alan invited me to speak at an American Fisheries Society symposium in Fairbanks. I had intended to leave for home in the afternoon of the meeting's last day, but the night before, I received word that my mother, who lived in San Diego, had been hospitalized, so I decided to fly directly from Fairbanks to San Diego. The flight rescheduling required that I spend another night in Fairbanks. Alan had been wanting to talk about some work he was doing, and the added time in Fairbanks provided an opportunity for us to do that.

Alan and I had an early breakfast together the morning of my departure, during which he showed me some data he had assembled on whaling in the North Pacific. I knew that an international ban on commercial whaling had been imposed in the late 1960s and early '70s, but I mistakenly thought that North Pacific whale stocks had been depleted many decades or even centuries earlier. I saw from Alan's data that, in fact, industrial whaling in the southern Bering Sea and North Pacific Ocean had not begun in earnest until after World War II. Alan posed a question to me that he had apparently been thinking about for some time: Were the great whales an important food resource for the region's transient killer whales? And if so, might their post–World War II depletion by industrial whaling have been the event that caused the transient killer whales to begin foraging more intensively on sea lions and sea otters?

I was staggered by Alan's proposal. My immediate visceral reaction to the idea was that it was too big and too simplistic. My immediate intellectual reaction to the idea was that it explained why sea lions declined when they did and thus resolved the mystery of a link between the two species' declines. In between attending to my mother's needs, I thought of little else over the next several days. I wrote to Alan and pointedly asked if he really believed the story he had told me. His brief answer—"No, but I don't disbelieve it either"—was just what I needed to realize that the idea was both worthy and in need of further exploration. A few months later, Alan came to Santa Cruz to work with me on doing just that.

The first step was to assemble a team to dissect the issues, pose more specific hypotheses, and test those hypotheses by analyzing the available data. The team's composition would be critical. Gus van Vliet, a long-time associate of Alan's who had been thinking about the interplay between industrial whaling and killer whales for years, was already a member. We approached Terrie Williams and Dan Doak, both of whom had worked with me on the sea otter–killer whale project, because we thought that the same kinds of demographic and energetic analyses we had used with the sea otters and killer whales would be useful for understanding the larger suite of interactions among killer whales, great whales, and various other killer whale prey. We also invited Eric Danner, who at the time was in the early stages of his doctoral program, to join us because he had both the time and the skills needed to delve more deeply into the International Whaling Commission's (IWC) database for the North Pacific.

As we discussed a way forward, Alan and I came to the realization that we should have an established member of the whale research community on our team, someone with an intimate knowledge of whale biology and someone who was respected by the international whale community. We needed such an individual to lessen the risk of misreading or misinterpreting the data, and to establish credibility with people who worked with great whales. Doug DeMaster, director of NOAA's National Marine Mammal Laboratory at the time and a well-known and respected figure in the marine mammal community, was an obvious choice. I had known Doug since graduate school, and I thought that if he were interested and willing, he would be ideal for the task at hand. A few weeks later, while at a meeting together in Honolulu, I laid out the general ideas and asked if he might like to join us. Doug seemed intrigued and agreed almost immediately.

Our team was rounded out with two others: Karin Forney, an employee of the NOAA lab in Santa Cruz and an expert on the population biology of small cetaceans; and Bete Pfister, a former UC Santa Cruz student and technician who would help Doug obtain and analyze records of marine mammal abundance in the North Pacific region.

From this point forward, our analyses and thinking moved quickly. First, we looked at the geography and timing of North Pacific harvest records in the IWC database. Large numbers of northern right whales and bowheads were harvested in the 1800s, as were humpbacks in the 1930s. But that was long ago, and the more recent slaughter was geographically and temporally aligned and, in aggregate, much greater in terms of numbers and biomass. Although the IWC database extended back to the early 1900s, few whales were commercially harvested from the North Pacific region in the decades before World War II. But at least several hundred thousand whales were removed from the region after the war. Early in that period, from 1946 through 1953, a modest number of animals (21,406) were harvested, nearly all of which came from three small areas off the coasts of Japan, Kamchatka, and British Columbia (figure 11.2). Whaling intensified over the next eight years (1954–62), spreading east across the North Pacific Ocean and southern Bering Sea and north along the western margin of the Bering Sea's continental shelf. The following eight-year period (1963–71) marked the pinnacle of North Pacific industrial whaling, with 92,403 recorded kills spread across the entire south-central Bering Sea and North Pacific Ocean. By 1972, industrial whaling in the North Pacific was well on the wane. The harvest during that period was 39,754 animals, most of which were sperm and Brydes whales in more

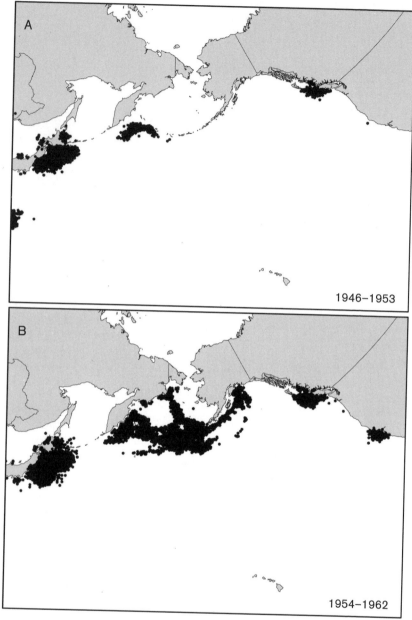

FIGURE 11.2. Locations of great whales harvested by industrial whalers from the end of World War II to the cessation of industrial whaling (from Springer et al., 2003; redrawn by Eric M. Danner).

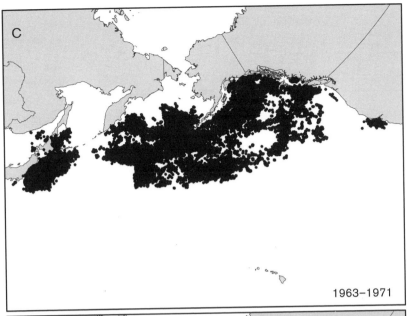

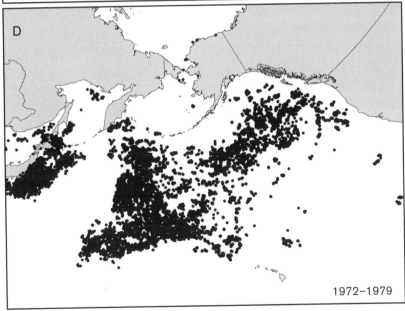

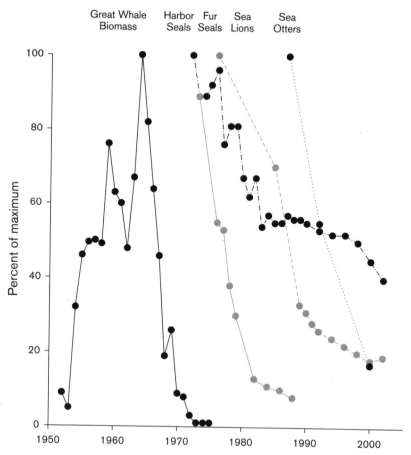

FIGURE 11.3. Relative biomass of great whales harvested through time within 200 nautical miles of the Aleutian archipelago, and the sequential collapse of pinniped and sea otter populations (from Springer et al., 2003).

southerly regions of the temperate to subtropical North Pacific. The post–World War II era of industrial whaling reduced whale biomass across the North Pacific Ocean and Bering Sea by an estimated sevenfold to eightfold.

Our interest was in the Aleutian Islands, so we plotted the number of reported kills within 200 nautical miles of the Aleutian archipelago and the coastal margin of the western Gulf of Alaska. Harvest numbers in this region rose sharply in the early 1950s, reached an apex in the early 1960s, and then collapsed to near zero by the early 1970s (figure 11.3).

Although our initial focus had been on sea otters and sea lions, two other pinniped species—harbor seals and northern fur seals—also occurred in the region and were known to have declined. The fur seal decline was complicated, partly because the population had been intensively harvested and partly because juvenile and female northern fur seals spend much of their lives far from the breeding rookeries of the Pribilof Islands. Harbor seal pups had also been exploited in the Gulf of Alaska. But harbor seals were more like sea lions and sea otters in that they had not been extensively harvested and did not venture far from haul-outs and breeding sites during nonbreeding seasons.

Harbor seals were abundant when I first visited Amchitka Island in the early 1970s, but population densities were notably decreased in my early period of study on Attu Island in the mid- to late 1970s. I didn't think much about the difference at the time, other than to imagine that it might have been caused by some unspecified difference between Amchitka and Attu. But as I continued working on Attu into the early 1980s, a declining population of harbor seals became more and more obvious. Low-lying islets in Massacre Bay, where I had commonly observed tens to hundreds of animals hauled out or swimming close to shore in the mid-1970s, were now nearly devoid of harbor seals.

I hadn't monitored the harbor seals of Attu because my work was focused on sea otters and on other parts of the food web that were more obviously linked to the sea otter–urchin–kelp trophic cascade. Nor had anyone else monitored harbor seal populations in southwest Alaska during those early years, except at Otter Island in the Pribilofs, a few sites in Bristol Bay, and Tugidak Island in the western Kodiak archipelago. At Tugidak the harbor seal population declined by almost 90 percent between 1973, when it was first counted, and 1980 (figure 11.3). Counts from Otter Island and Bristol Bay, while spottier, showed similar patterns. These data matched what I had seen in the Aleutians and what a few others had seen in areas to the east, so I assumed that they provided a fair characterization of the timing and magnitude of the harbor seal decline.

When lined up together, these various data revealed a startling pattern: a sharp rise and fall of whaling landings from the early 1950s through the late 1960s, followed by the sequential decline of harbor seals, Steller sea lions, and sea otters (figure 11.3). The pinniped decline and that of sea otters, it seemed, must be linked by some common thread. In my mind, the evidence weighed fairly strongly against that thread originating at the base of the food web and acting on these various species through bottom-up forcing. For one, the otters and pinnipeds were nourished by different ecosystems—coastal benthic production in the otter's case and offshore water-column production in the case of the pinnipeds. The weight of evidence from sea lions argued

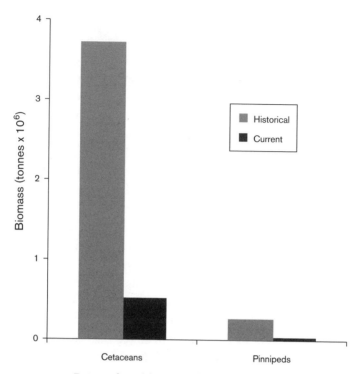

FIGURE 11.4. Estimated total biomass of cetaceans and pinnipeds before (light gray) and after (dark gray) the era of industrial whaling in the North Pacific Ocean (from Springer et al., 2003).

against food limitation, and the weight of additional evidence from sea otters pointed to killer whale predation as the likely proximate cause of the population decline. The magnitudes and rates of population decline were broadly similar for each of the three species. Armed with these insights and information, it was an easy next step to suspect that all three species collapsed because of increased predation by killer whales.

A final piece of information that would become important to our thinking was an analysis based on the abundances of killer whales and their prey. These data, together with estimates of the field metabolic rate and assimilation efficiency of killer whales and the vital rates and growth rates of prey and their caloric values, allowed us to assess the sustainability of killer whales and their prey under a variety of dietary scenarios (Estes et al., 2009a). Estimates of prey abundance and per capita biomass revealed that, prior to the megafaunal collapse, whale biomass exceeded that of the pinnipeds and sea otters by one to two orders of magnitude (figure 11.4). This difference alone suggested that a transient killer whale population that had grown or evolved to consume great whales, and whose abundance was set by the secondary production of great whales, could not be sustained on a diet of pinnipeds and sea otters alone. We tested that hypothesis by assuming a transient killer whale population of about 400 individuals (the approximate number of identified and therefore known individuals in modern times) and then calculating whether these killer whales could be sustained on various dietary scenarios in which differing species, proportions of total deaths, and proportions of each dead carcass were consumed by them. Inasmuch as our question was whether the reduction of great whales caused killer whales to turn to and deplete populations of smaller marine mammals, we conducted the analysis by dividing the marine mammal prey into two groups: great whales (blue, Brydes, fin, gray, minke, sei, humpback, bowhead, northern right, and sperm) and smaller marine mammals (harbor seal, harbor porpoise, sea otter, beluga whale, northern fur seal, Steller sea lion, and Dall's porpoise).

These analyses indicated that a population of 400 transient killer whales was potentially sustainable on a diet that did not include large whales, but only if the killer whales consumed all of each carcass from every death event in all the smaller species of marine mammals. While theoretically possible, this seemed highly unlikely. We concluded, therefore, that great whales must have been consumed in some considerable number and amount to sustain the transient killer whale population, and thus that the reduction of great whales by industrial whaling would have caused the transient killer whales to die, move elsewhere, or increase their consumption of smaller marine mammals at unsustainably high rates.

Although uncertainties remained about both data quality and mechanism, our interpretation of what must have happened was straightforward and obvious. Before industrial whaling, the transient killer whales were sustained by feeding on the immense biomass of great whales in the North Pacific Ocean and southern Bering Sea. Whether killer whales fed on the smaller marine species during that period is unknown, but if the consumption of great whales provided a sufficiently high net rate of energy gain, then foraging theory suggested they did not.

With the reduction in abundance of great whales through industrial whaling, the rules of consumer choice changed; the net rate of energy gain to killer whales from great whales declined and killer whales thus expanded their diet to include smaller and less rewarding pinnipeds and sea otters. We postulated that these secondary prey were unable to sustain the increased mortality rates associated with killer whale predation, and hence their populations collapsed, one after the next, as the killer whales continued to expand their diet to include increasingly less profitable prey. This scenario, published in *Proceedings of the National Academy of Sciences* in October 2003 (Springer et al., 2003), became known as the "megafaunal collapse hypothesis."

The underlying logic of the megafaunal collapse hypothesis rests, to some significant degree, on the economics of consumer choice—that is, the assumption that the net rate of energy gain to killer whales from great whales is higher than it is from the smaller marine mammals. Great whales are the largest and most calorically rich organisms in all the world's oceans. Any consumer capable of feeding on this prey resource ought to do so. But the net rate of energy gain also must account for the costs of capture and assimilation, which can be great with large and dangerous prey like a great whale. For this reason, it remained unclear whether the net benefits of feeding on whales would surpass or fall below those of feeding on smaller marine mammals, which might be captured and subdued with much less risk and effort.

If transient killer whales derived much of their nutrition from the great whales, how did they do this in a suitably efficient manner, in a way that allowed them to realize a higher net rate of energy gain than they could get from the smaller marine mammals? Although the information required to calculate and contrast net rates of energy gain from each member of the transient killer whales' potential prey base of marine mammals was unavailable, subduing a living whale is clearly a formidable challenge. For this reason, transient killer whales often target great whale calves, and we imagined they might also target weakened older individuals at or near the ends of their natural lives.

Hal Whitehead and Randy Reeves—two people intimately familiar with the natural history of great whales and the history of whaling—added an interesting twist to these ideas (Whitehead and Reeves, 2005). Whitehead and Reeves pointed out that the process by which whalers killed and retrieved whales would have both advertised the moment of death and created an undefended and easily consumable resource for transient killer whales. The "advertisement" occurred via the noise produced by an exploding harpoon, which presumably would have been detectable by transient killer whales at some distance. The killer whales might have learned to associate these sounds with food, just as certain resident killer whales and sperm whales have learned to associate sounds produced by the hydraulics of fishing vessels with the retrieval of deep-water long lines, to which they are attracted and from which they commonly pilfer the catch before it can be brought on board.

Whitehead and Reeves pointed out further that freshly killed whales were injected with air to keep them afloat until larger processing vessels arrived and brought the whale carcass aboard for rendering. Many hours commonly ensued after a whale was killed, fluked, and floated by the catcher ships and before it was drawn aboard for processing by the larger factory ship, during which time killer whales often scavenged the carcasses.

The depletion of living whales may have been sufficient to precipitate a change in killer whales' foraging behavior. The scavenging hypothesis of Whitehead and Reeves identifies a related process that could have contributed significantly to this behavioral change. The sudden provisioning of an energy-rich food resource, for which both the search time and the cost of pursuit and capture were greatly reduced or eliminated, almost certainly increased the net rate of energy gain from great whales to transient killer whales beyond that which had been realized from anything previously available to them. Over the roughly two decades of intensive industrial whaling around the Aleutian Islands, this may have enhanced the local population of transient killer whales and perhaps even caused individuals that had previously fed on smaller marine mammals to forgo their more traditional prey in favor of great whale carcasses. In the beginning, this would have reduced killer whale–induced mortality on smaller marine mammals by the degree to which that mortality was additive, thus causing populations of these smaller marine mammals to increase by a commensurate amount.

This chain of events might explain why sea lions and harbor seals were so abundant across southwest Alaska in the early 1970s. All of that would have ended abruptly in the late 1960s and early '70s with the cessation of industrial whaling. Not only had great whale populations been decimated, but the whale carcasses for scavenging suddenly disappeared as well. Any transient killer whales that had been feeding on and benefiting from this prey resource were inevitably forced to look elsewhere for food. If even just a few of these animals turned to smaller marine mammals, this would explain why the pinniped declines began when they did and why their onset was so abrupt.

The megafaunal collapse hypothesis was sexy because of the charismatic players, and appealing because it explained the main patterns of change in marine mammal abundance in southwest Alaska over the past half-century. But was it right? We couldn't be sure and tried to reflect this uncertainty with appropriate caveats alongside the arguments. To avoid any misunderstanding of intent, we ended the paper's title with a question mark: "Sequential megafaunal collapse in the North Pacific Ocean: an ongoing legacy of industrial whaling?" We published the paper not because we believed that the megafaunal collapse hypothesis was true, but because it seemed to line up with the available data better than the widely accepted dogma of bottom-up forcing. Our intent was to convey the same sense of uncertainty that had been in Alan Springer's answer to my question a few years earlier when I asked if he believed that the collapse of pinnipeds and sea otters was ultimately linked to the effects of industrial whaling—"No, but I don't disbelieve it either." We thought that the megfaunal hypothesis might be true, we thought the available data lined up with this explanation better than those that had been proposed earlier, and we hoped the paper would motivate others to rethink or at least expand both their beliefs and the directions of their future research activities.

I expected and hoped that the megafaunal collapse hypothesis would be received much as the killer whale–sea otter hypothesis had been received five years earlier—with interest and an appropriate level of skepticism. I could hardly have been more naive. Although the paper attracted a great deal of media attention, it also created a firestorm of controversy and conflict in the marine mammal community.

TWELVE

Whale Wars

THE MEGAFAUNAL COLLAPSE HYPOTHESIS was a sweeping challenge to conventional thinking about whale and pinniped ecology. My colleagues and I expected it to be met with skepticism. But we also thought it would promote interest and discussion. We were prepared for tough questions. We were wholly unprepared for the anger and vitriol that ensued. This chapter is my account of what happened, with reflections on why.

All of the key ideas and most of the supporting evidence for the megafaunal collapse hypothesis were in place by 2000, at which point we began thinking about how best to present the hypothesis to others. The first step was to talk about it, which for me occurred largely via seminars and presentations at scientific meetings. I gave one such presentation at the 2000 meetings of the American Association for the Advancement of Science (AAAS) in Boston as part of a symposium organized by Jeremy Jackson on marine historical ecology. The following evening, at a cocktail party, I was introduced to Don Kennedy, editor-in-chief of *Science*. Kennedy had not heard my talk but was given a brief synopsis during our introduction. The hypothesis must have struck him as interesting, because after I recounted the story, he asked if I would consider sending a report on the work to *Science*. I reminded him of the story's empirical shortcomings and hypothetical nature, but he still seemed to want us to submit something and suggested that we write it as a "Perspective," which would provide more latitude for speculation and uncertainty. In parting, Kennedy reiterated his interest and left me with the impression that he would "grease the skids" (my words, not his) by speaking with the subject editor for the Perspectives section. This seemed like a stroke of terrific luck.

That was in January. By February we had a manuscript written and off to *Science*. The Perspectives editor responded quickly to confirm receipt, express further interest in publishing our paper, and request some changes in formatting and presentation. We were encouraged by the positive reaction and returned a revised manuscript to *Science* later that month. Initially, I was almost certain the paper would be published. But as the months passed with no word back from the journal, I began to think something was amiss. Then, in July, I received a standard form letter from *Science* rejecting the article. Attached to the letter were several reviews, one of which proclaimed nutritional limitation as the cause of the Steller sea lion's decline, thus concluding that our megafaunal collapse hypothesis must be wrong. This was a setback, or so it seemed at the time.

As my colleagues and I worked through the data for the megafaunal collapse hypothesis, we began to think more widely about whale ecology and the consequences of whaling on ocean ecosystems. These interests morphed into a workshop proposal to explore the issue. The idea was to invite a mix of people, including whale experts and experts on ecological process. The workshop was conceived with three processes in mind: great whales as consumers (the ecological roles of whales through impacts on their prey populations); great whales as prey (the ecological roles of whales through impacts on the creatures that eat them, such as killer whales and sharks); and great whales as detritus (the effects of whale falls on seafloor ecosystems). I organized the conference in collaboration with four colleagues—Dan Doak and Terrie Williams at UC Santa Cruz, and Doug DeMaster and Bob Brownell with NOAA—and arranged to publish the proceedings through the University of California Press. It was an exciting time; the potential for moving our understanding of whale ecology forward seemed immense.

The workshop was held in Santa Cruz at the Chaminade Resort over four days in April 2002. The talks were interesting, but the interactions between whale biologists and ecologists were more strained than I had hoped they would be. I sensed a divide between those who studied whales and those who did not. The ecologists brought theories and findings from work on other species in other ecosystems. Many of the whale biologists were skeptical and even closed-minded toward these theories and results, or so it seemed to me. I came away from the workshop with the suspicion that at least some of the whale biologists were not there to learn about whale ecology, but rather to better understand what the ecologists had up their sleeves and to keep them honest. Before departing, several of the ecologists privately expressed dismay over what they took to be hostile anti-intellectualism by the whale community.

Although the workshop failed to achieve the intellectual integration and synthesis I had hoped for, it succeeded in leading to an edited volume (Estes et al., 2006) and helping me understand, in small measure, what we were up against with the megafaunal collapse hypothesis. Although I didn't see it at the time, a battle line on this issue had already been firmly drawn.

Bob Paine was a workshop attendee, and he had watched the megafaunal collapse hypothesis unfold through one or two of my earlier talks, more detailed discussions of the ideas and data that transpired at meetings of the National Research Council's sea lion study group, and probably through independent discussions with one or more of my collaborators. He also knew of our failed submission to *Science* and seemed to feel, as I did, that we had been unfairly treated. Bob also had come to suspect that the sea lion community was on the wrong track with nutritional limitation and thought that the megafaunal collapse hypothesis needed to be on the table for these people and others to think about. He suggested that we revise the failed *Science* manuscript and send it on to him for consideration by *Proceedings of the National Academy of Sciences (PNAS)*. There was no guarantee of publication, but Bob could assure us a fair and impartial review. His words to me were something like "I can at least be sure they don't send it to someone like that guy in the back of the room."

We revised the paper and then asked several colleagues to read it for content, argument, and tone. One of these "friendly reviewers" was Mary Power, my friend and respected colleague at UC Berkeley, who offered the simple but brilliant suggestion that we end the paper's title with a question mark. My coauthors and I had been sensitive from the outset to the speculative nature of the megafaunal collapse hypothesis and had struggled mightily in our search for ways to modulate the story with just the right mix of argument and caveat. Mary's proposal to include the question mark did that perfectly, as an unequivocal sign of uncertainty at the very beginning. This, we thought, would allay any danger of a reader misinterpreting our paper as a statement of belief or fact as opposed to the proposal of a hypothesis. The formal reviews were critical but fair, and overall they supported publication. After some back-and-forths with further revisions, the paper was accepted for publication in *PNAS*.

Formal acceptance of our paper was followed by the first of many unpleasant surprises. Doug DeMaster wrote to ask that his name be removed from the byline. Doug's rationale for withdrawing was twofold: first, that it would be inexpedient for him, as chair of the Scientific Committee for the

International Whaling Commission, to have his name on such a potentially controversial paper; and second, that the paper had become more of an argument than a hypothesis. Of course, we had no choice but to honor his request. But neither point rang true with me. If authorship were politically sensitive, he should have figured that out earlier, before we had submitted the manuscript to *PNAS*—and even before we had submitted the original version to *Science*. If anything, I thought that we had softened the paper both in overall tone and with the interrogative title. As it was, Bob Paine was so upset with Doug's withdrawal that I worried he might intercede and prevent the manuscript from being published. Bob spoke with Doug, and apparently Doug convinced him that his reasons for withdrawal were sound. More importantly, though, we had lost an important colleague and ally.

The megafaunal collapse hypothesis appeared in *PNAS* in October 2003 (Springer et al., 2003), followed immediately by an outpouring of media coverage. Recountings of the story were everywhere—in the newspapers, on radio and television, and, later, in more detailed written summaries. Unfortunately, as nearly always happens, important nuances were lost to the media. What was clearly stated as a hypothesis in our published paper became a simple fact. And that didn't set well with the skeptics, especially those who didn't believe the hypothesis in the first place. A groundswell of negative response developed rapidly, leading to the proposal for a plenary point–counterpoint discussion and debate at the upcoming biennial conference of the Marine Mammal Society in Greensboro, North Carolina, later that fall. Paul Wade from NOAA emerged as the leading dissenter and pulled together a long list of supporters, many of whom I barely knew or had never heard of. Terrie Williams and I agreed to represent our side while Paul would present the counterpoint. We would have an hour, first 12 minutes each from me and then Terrie, followed by 25 minutes from Paul. I would then have five minutes at the end to respond to whatever Paul might say.

The lead-up to the biennial conference was unnerving. I knew the many soft points to our hypothesis, but it would be difficult for Terrie and I to convey the hypothesis, its strengths and weaknesses, and our views on the flaws in earlier thinking in just 24 minutes. In addition, I had no idea what information Paul might have or what he would say. However, my sense from his abstract and what I had learned through the grapevine was that Paul and his associates knew we were wrong and could prove it. I didn't know how that could be, but I was concerned nonetheless. No one relishes the possibility of being publicly embarrassed at a large international meeting. But the purpose of the point–counterpoint was to put our views and supporting information on the table so that the listeners could draw their own conclusions. I felt that we were at a great disadvantage, because while Paul knew the details of our data and arguments, we knew nothing of his. Going into the Greensboro meeting, I had resigned myself to acknowledging whatever substantive points came from Paul's presentation and, at the end, to modifying my position accordingly. After all, the goal of science should be a search for truth.

The lecture hall was electric with anticipation of our debate. Terrie and I made our points in the time allotted. As I sat down and prepared myself for what Paul might say, I was already thinking of how to incorporate the conflicting evidence he would provide into my five-minute wrap-up. But as Paul's talk drew to a close, I realized that he had said very little, other than that ecosystems are complex and the data were thinner than he would like. He didn't explain why we were wrong, and he didn't propose an alternative explanation for the species declines to challenge our megafaunal collapse hypothesis.

I thought Paul's presentation was weak, and I said as much when I took the podium to close the session. Specifically, I accused him and his supporters of two fundamental transgressions in a challenge to any scientific argument—failing to provide and test alternative hypotheses, and hiding behind what I called "the tyranny of Type I error." Type I error, in statistical lingo, is a failure to reject the null hypothesis of no effect (in our case, no effect of the purported megafaunal collapse hypothesis) when some alternative hypothesis is in fact true. The "tyranny" of this has to do with making the standards of inference so rigid and extreme that the null hypothesis could never be rejected with the sorts of data that were available. And I was annoyed at Paul for making such a kerfuffle with so little real ammunition. Several nights later Paul approached me in the hotel bar to express his displeasure. I apologized for hurting his feelings but then told him that I had simply said what I believed to be true.

Debates at meetings and conferences have little substance because they often turn on a speaker's intrinsic ability to persuade, never allow a thorough exploration of the issues, and almost always are forgotten with time. I thought it important to move beyond that and commit the arguments in writing to the peer-reviewed literature, thereby providing, in perpetuity, the information needed by any interested party to carefully explore the ideas and issues. I thus approached Don Bowen, editor of *Marine Mammal Science,* about publishing such an exchange.

We had already published the megafaunal collapse hypothesis. The proposal was for Don to invite Paul Wade and colleagues (and, if he wished, others with independent views) to write rejoinders; for Alan Springer and colleagues to reply; and for the exchanges to be published together in the journal. Don agreed, and it all started well enough with a plan for separate rejoinders by Paul Wade's group and Andrew Trites's group, and a response to both by Alan and his colleagues. The Trites manuscript was quickly written and submitted but the Wade manuscript languished. In the meantime, I took over as the journal's editor when Bowen stepped down. I asked Bowen to see the exchange through to publication, but he declined. This left me in the awkward position of being both the editor of the journal and a party to the debate. I chose to resolve this conflict of interest by recusing myself from further responsibility over the exchange and asking Ward Testa, one of the journal's subject editors, to see the papers through to publication. I chose Ward because as an employee of NOAA and the National Marine Mammal Lab, I thought he understood the issues and I thought his leadership would allay any concerns of the opposing parties about fair and objective treatment in the journal.

The Wade manuscript remained in limbo long after the Trites manuscript had been accepted, and Trites began to press Testa to publish his paper before the issue was forgotten. By the time Wade's manuscript, which was extremely long, was finally received, pressure from Trites was extreme and Testa pushed Springer to quickly write a response. Springer resisted, claiming the need for reasonable time to prepare a proper retort. Testa responded by threatening to publish the various papers in different journal issues if Springer did not submit his manuscript in what Testa thought was a timely manner. The exchange between Testa and Springer turned ugly as Testa further announced his intent of possibly allowing Wade to have the last word on the issue.

As editor-in-chief of the journal, I was uncomfortable with Testa's intentions. I had given him authority to review, edit, accept or reject, and eventually publish the manuscripts together in a single issue of the journal. In my mind, he did not have the authority to give Paul Wade a final word in response to Springer or to spread the papers out into different issues.

The Springer manuscript was eventually completed and submitted, but not soon enough to be slated for publication together with the Wade and Trites manuscripts in the next issue of the journal. I was appalled, and I stepped in by instructing my editorial assistant to hold publication until the three papers could be published together. Testa wrote the Marine Mammal

Society's Board of Governors, charging me with editorial misconduct. I responded by recounting the history of the exchange and my agreements with Testa, stating my belief that Testa had not been fair and objective in his handling of the matter and requesting that the board intercede by removing Testa from any further editorial responsibility and replacing him with someone else of their choosing who would see the papers through to publication, as originally intended. The Board of Governors discussed the issue and chose to allow Testa to see the papers through to publication. In the meantime, Testa rejected Springer's paper, thus rendering moot any possibility of publishing the papers together. I took the Board of Governors' decision as a vote of no confidence in me as their journal's editor and stepped down. Meanwhile, the Trites and Wade papers were published (Trites et al., 2007; Wade et al., 2007).

Springer's response (Springer et al., 2008) was eventually accepted for publication by Daryl Boness, the journal's new editor, but not before blood had been shed. And the bloodshed didn't end there. Wade was unhappy with Springer's response and submitted yet another rejoinder. Boness was willing to publish Wade's rejoinder but only after allowing us to respond and with the understanding that that would be the end of any further published exchanges on the issue in *Marine Mammal Science*. These papers were published together in 2009 (Estes et al., 2009b; Wade et al., 2009).

In the meantime, Doug DeMaster and several others published an independent critique of the megafaunal collapse hypothesis (DeMaster et al., 2006). I could understand and live with Doug's decision to withdraw as a coauthor on the *PNAS* paper, but his decision to then turn around and publish a critique troubled me. Doug and I had been friends since just after graduate school. I liked and respected him as much as anyone in marine mammal science. I couldn't imagine how his view of the science had changed so radically, especially considering the lack of any new information, and thus could only conclude that his decision to write the critique was motivated by political or economic pressure.

The floodgates were now open, and others moved in to pick at the bones of the megafaunal collapse hypothesis. Sally Mizrock and Dale Rice, both from NOAA's National Marine Mammal Laboratory, rejected the hypothesis on the basis of a purported lack of large whales in killer whales' diet (Mizroch and Rice, 2006). Some years later, in a paper published by *Mammal Review,* Katie Kuker and Lance Barrett-Lennard returned to the source by challenging the evidence that the sea otter decline had been caused by killer whale predation (Kuker and Barrett-Lennard, 2010). I was not invited to reply to Kuker and Barrett-Lennard, and when my colleagues and I attempted to respond to what we knew was flawed reasoning and bad scholarship by sending a rejoinder to *Mammal Review,* our paper was rejected.

I've recounted the history of the debate over the megafaunal collapse hypothesis, but what were the issues? In fact, there was little of substance, at least in my view. In the remainder of this chapter I'll outline the key points of scientific contention, briefly summarize our response to those points, and speculate on why the megafaunal collapse hypothesis provoked so much push-back. The critiques and arguments are all in the literature, and I would encourage anyone whose curiosity has been piqued by my account here to read these papers in the order they were published. The papers are available online at www.ucpress.edu/go/serendipity.

Published criticisms of the megafaunal collapse hypothesis revolved around the following five observations or arguments:

1. Killer whales seldom attack or eat great whales.
2. The collapses of the pinniped and sea otter populations were not sequential.
3. Nutritional limitation caused or contributed to the pinniped declines.
4. The declines of great whales and the collapse of pinniped and sea otter populations in southwest Alaska do not properly line up in time.
5. The geographic extents of whaling and the pinniped–sea otter population declines do not properly line up in space.

Our critics saw these points as flaws in the megafaunal collapse hypothesis, but we disagreed on both empirical and logical grounds.

The argument that killer whales seldom attack or eat great whales was, in our view, simply wrong. There had been numerous observations of such attacks, and a large proportion of living great whales bear rake marks from killer whale teeth. Moreover, written and artistic records of the whalers themselves indicate that this was a common occurrence in earlier times. There have been hundreds, perhaps even thousands, of sightings of attacks by killer whales on great whales worldwide, though most of them not by scientists.

Arguments for the sequential nature of the collapses of pinnipeds and sea otters were based on the critics' inability to demonstrate significant differences in the timing of the various population declines. I still see this argument as nothing more than statistical tomfoolery, with the critics failing to establish test criteria that balanced the probabilities of both Type I error (inferring that the decline was sequential when in fact it was not) and Type II error (inferring that the decline was not sequential when in fact it was). The weight of available evidence for a sequential decline, in my view, was reasonably strong (see figure 11.3; see also Estes et al. 2009b).

I was frankly amazed by the critics' adherence to a belief in nutritional limitation. My own perhaps overly cynical view is that people see and believe what they think is true. And many ocean scientists take it almost as fact that population fluctuations result from bottom-up forcing. Food is indeed an essential resource for all consumers. When food is scarce or nutritionally inadequate, populations of consumers that depend on that food for sustenance will decline. Pinnipeds and sea otters in southwest Alaska may well have experienced some nutritional limitation at times and in places, but the overall weight of evidence is simply inconsistent with nutritional limitation as the principal driver of the population declines.

Some of our critics charged that the depletion of great whale numbers and biomass in the North Pacific Ocean and southern Bering Sea occurred much earlier than the megafaunal collapse hypothesis implied. But the IWC data do not support that contention (the data in figures 11.2 and 11.3 make the point about as clearly as possible).

A final general point of argument by some of our critics suggested a geographic inconsistency between the extent of industrial whaling and the collapses of sea otter and pinniped populations. More specifically, they asked why, if the megafaunal collapse hypothesis were correct, it didn't lead to the same outcomes in places like southeast Alaska and British Columbia? The same question was asked of the Commander Islands and Russia's east coast.

This is the one point that gave me pause to wonder. Perhaps the lower pre-exploitation abundance and biomass of great whales in these latter regions caused killer whales to depend less on great whales for food? Or it might be that pinniped and sea otter numbers in these regions were insufficient to draw the killer whales' attention to them as strongly as in the Aleutians. However, the fact remains that killer whales do prey on pinnipeds and sea otters in southeast Alaska and British Columbia. I don't believe that these various assertions invalidate the megafaunal collapse hypothesis, but I'm unable to explain with any reasonable confidence why the sea otter and pinniped collapses were limited to southwest Alaska.

I have pondered why the megafaunal collapse hypothesis was met with such resistance. It can hardly be because of the strength of contrary or inconsistent evidence, for there is little or none of that. And except for Trites's unwavering belief in the late-1970s regime shift in the Pacific Decadal Oscillation as the cause of the sea lion's decline, which itself has important shortcomings (see chapter 11), there have been no alternative explanations for it, let alone hypotheses that tie together this and all the other declines observed in southwest Alaska during the same period. Part of the resistance might stem from differences in philosophy and belief about why ecosystems change and the nature of evidence required to explain those changes. Some of our critics undoubtedly viewed the megafaunal collapse hypothesis as oversimplistic, perhaps believing that such profound change in any complex system is never attributable to a single factor. Ecosystems are admittedly complex, but experimental ecology has repeatedly shown that single factors often have strong effects on the structure and organization of ecosystems.

The nature of evidence is another matter. No small number of scientists familiar with the system and the issues, including some of my friends and allies, believe that supporting evidence for the megafaunal collapse hypothesis is too thin to warrant a scientific argument. My colleagues and I were acutely sensitive to this issue, which is why we ended the title of our initial paper with a question mark. I find it difficult to believe that this latter point of view was at the heart of the push-back, given the flimsy nature of the arguments and supporting evidence against the megafaunal collapse hypothesis.

I have thus come to suspect that the push-back was largely rooted in the darker corners of science, namely those shaped by economics, policies, and human behavior. Some of it may have been the result of adverse reactions to an explanation for the pinniped declines on the part of people who had tried to solve the problem but had failed. Part of it may have been a fear for the policy implications—that the megafaunal collapse hypothesis would absolve the ground fishery of ecological wrongdoing or even lead to the management of killer whales. And some of it may have been for worry by the established research community over what it would mean for their future funding. Hundreds of millions of dollars had been spent on sea lion research, ostensibly in search of explanations for why the populations had declined and why they weren't recovering. The sole justification for these vast expenditures was the sea lion's threat to the valuable ground fishery via legal intervention under the *Endangered Species Act*. So long as the reason for the sea lion's collapse was unknown and the ground fishery was potentially culpable, funding would continue. But if the megafaunal hypothesis were correct, funding for sea lion research would almost certainly diminish, and with it people's research programs and even their jobs. I'm sure this was a factor in the push-back because I heard a number of people say so, once even in public at a scientific meeting.

Conflicts are never black and white. Right or wrong in science is always, to some degree, a matter of opinion and perspective. I've provided my perspective, but how and why might someone on the other side see it differently? One might ask why we published the megafaunal collapse hypothesis in the first place, given all the uncertainty and the inevitable trouble it would create. Part of it was simply our backgrounds and motivations, which differed from those of our critics. We were mostly academic scientists, and success in academia stems from creativity and discovery. We were more inclined to speculate than most of our critics, in part because that's what academics do and in part because we didn't have to live with the economic and policy consequences. But part of it was also frustration over the established research community's failure to understand the pinniped declines and the lack of ideas over how to move forward. The megafaunal collapse hypothesis was intended to be as much a nudge to these people as it was an explanation for what really happened.

If I have one regret, it is over my behavior as editor-in-chief of *Marine Mammal Science*. I should have recognized that a fair and proper airing of the conflict over the megafaunal collapse hypothesis was unachievable in a journal for which I was also the editor, and I should have sought to have the exchange published elsewhere. Otherwise, I have few misgivings and probably would act much as I did if I could rerun the clock.

THIRTEEN

Foxes and Seabirds

MY WORLDVIEW OF ECOLOGY had expanded by the beginning of the twenty-first century to include two fundamental beliefs: that species were linked in complex ways within and between ecosystems, and that predators and top-down forcing processes nearly always contributed to the linkages. There was nothing revolutionary in these beliefs. Many other ecologists probably see it in more or less the same way. But I had the advantage of vivid, empirically based images from my wanderings among the islands of the Aleutian archipelago and watching them change over the decades. I had seen how the influence of sea otter predation spread through coastal ecosystems to affect kelp, mussels, barnacles, fish, gulls, and sea stars, and I had grown to suspect that many or most other species in the ecosystem were similarly affected (see chapter 8). I had also witnessed two events that involved linkages between the coastal oceans and the open sea, one being the episodic influx of smooth lumpsuckers (with their subsidizing influence on the sea otter population; Watt et al., 2000) and the other being the arrival of killer whales (with their depleting influence on the sea otter population, and all that followed; Estes et al., 1998). Certain details of how and why these two events occurred as they did were arguable. Yet there was no denying their existence or that they served to connect the coastal ocean and open sea in ways that strongly influenced coastal ecosystems.

Evidence of trophic cascades was emerging from diverse ecosystems (Pace et al., 1999), which not only established a controlling influence of predators on plants but implicated more far-reaching linkages within food webs, through bottom-up effects of altered plant communities on myriad species and ecological processes. Some ecologists still viewed species linkages within ecosystems from a strongly bottom-up perspective, but others were beginning to understand that food-web dynamics operated through a more complex mixture of, and interplay between, bottom-up and top-down control.

Linkages across ecosystems were more difficult to see and understand, in part because they were less amenable to experimental study and in part because ecologists hadn't thought to look for them. The late Gary Polis probably did more to rectify this situation than any other single ecologist. Gary's views and arguments for the importance of cross-ecosystem linkages grew from his work on islands in the Gulf of California (Polis and Hurd, 1996), where he discovered that marine-derived nutrients subsidized terrestrial production, especially during wet years, when water was not limiting. Others began to look at nature with a similar view. In the Pacific Northwest, stream and forest ecologists saw that the spawning migrations of salmon transported nutrients from a productive ocean to intrinsically impoverished streams, from which point those nutrients were transported by predators and scavengers to the surrounding land, where they significantly enhanced plant growth (Naiman et al., 2002). Although independent studies were beginning to look for and discover similar intersystem linkages elsewhere in nature, the most compelling evidence came from juxtaposed aquatic and terrestrial systems, because critical limiting resources often varied strongly between these systems and because the ecosystem interfaces were sharply defined and thus easy to see and study.

I hadn't thought very much about the ecological importance of land–sea connectivity until the early 1990s. Gary's work opened my mind to the process, and the salmon–forest connection was both intriguing and compelling. Through all of this, my interest in and view of process in the Aleutians were focused exclusively on the marine realm. Land, for me, was simply a place to live and walk around while working on and in the ocean. I vaguely wondered whether the sea otter's ecological influence reached into the terrestrial realm through the windrows of kelp wrack that washed ashore, but I hadn't noticed differences in the terrestrial plant communities between islands with and without otters, and I long gave the matter no further thought.

That changed after a chance reunion with Vernon Byrd in the mid-1990s (figure 13.1). I met Vernon in 1971, while I was still a graduate student and he was a navy ensign stationed on Adak. Vernon was also a biologist and had become intensely interested in the Aleutians, especially its avifauna. This interest rapidly grew into a love of the place, and so after his naval discharge Vernon signed on as a biologist with the Aleutian Islands Unit of the Alaska Maritime National Wildlife Refuge. Although his responsibilities were initially focused

FIGURE 13.1. Vernon Byrd, recording data from a goose nest in the Aleutian Islands. (Photo by Jeff Williams)

on the endangered Aleutian cackling goose, he saw the Aleutians as a seabird refuge and as a place to monitor and learn about seabirds.

The refuge owned and operated the M/V *Tiglax* (pronounced "Teck-la," the Aleut word for "eagle"), which they used to support field activities in this remote and otherwise inaccessible region. Our meeting on Adak in the mid-1990s occurred as Vernon was ending a cruise on the *Tiglax* and I was beginning one. Vernon's field crew and equipment had been off-loaded and mine were already on board. We were scheduled to leave early the next day for Amchitka and several other islands to the west.

Vernon and I hadn't spoken at length in years, so we took the evening to catch up. As our musings wandered through the things we had done and learned, talk turned to the land and Vernon's reflections on how difficult it was to walk around on Buldir Island now that he was getting older. Although I had never been on Buldir, this struck me as odd because I found walking across the Aleutian tundra to be fairly easy, except along shoreline margins where a narrow band of tall grass and dwarf shrubs was indeed difficult to traverse. I puzzled over this, as Vernon and I were about the same age and he seemed as fit as I was. Buldir was mountainous, but then so was Attu. I didn't understand the difference, but Vernon did. The difference, he explained to me, probably had to do with the interplay between nutrients, seabirds, and introduced arctic foxes.

Because they are oceanic islands, the only vertebrates to colonize the Aleutians were those that could swim or fly. This meant that terrestrial mammalian predators were lacking—or had been until humans put them there after the sea otters' demise from overhunting and the resulting collapse of the maritime fur trade. Arctic foxes were introduced to nearly all of Alaska's larger islands, including most of the Aleutians, in an effort to maintain a flow of revenue from the sale of their winter pelts (Bailey, 1995). Before that, the Aleutians supported vast numbers of breeding seabirds whose populations were maintained by the rich surrounding ocean. Introduced foxes decimated the seabirds, which as ground-nesters on treeless islands were defenseless against the exotic predators.

Similar accounts of population collapses in island seabirds were known from the introductions of cats, mongooses, snakes, and various other predators to what had previously been predator-free islands around the world. But little thought seemed to have been given to the consequences of the seabirds' declines. Vernon proposed that the comings and goings of seabirds between land and sea fertilized the islands with guano, unconsumed food, and their own dead bodies, thus enhancing terrestrial plant growth in an environment where nutrients were otherwise limiting. This hypothesis rested largely on the assumption of intrinsic nutrient limitation, which, at the time, I had no reason to think was either right or wrong. As we discussed that particular point in greater depth, Vernon recounted some of the things he had seen to make him suspect that nutrients were indeed intrinsically limiting to Aleutian plant communities. Vernon was a Georgia boy who could tell a story with the best of them. He described the telltale signs of caribou kill sites on Adak and ancient Aleut midden sites throughout the Aleutians by their earlier-than-normal green-up in the spring. With my focus on the ocean, I had never thought to look for such things. But the following spring, while working in the Shumagin Islands off the southwest coast of the Alaska Peninsula, I saw the same early green-up of a narrow band on the islands' seaward margins and around the nest sites of bald eagles farther inland.

I was intrigued but still not convinced by the hypothesized links among foxes, seabirds, and terrestrial plant communities. The idea made sense and seemed to be well supported by preliminary observations, but further study would be needed to put the hypothesis to a rigorous test. I had almost no experience in terrestrial plant ecology, and my first inclination was to leave the work for someone who was better versed in that arena. But the idea continued to intrigue me, and as my other fieldwork in the Aleutians was wrapping up, I began to think more seriously about taking it on. The hypothesis also had strong conceptual appeal, in part because it involved both inter-ecosystem linkages and a mix of bottom-up and top-down control, and in part because it potentially brought another remarkable dimension to the sea otter's sphere of influence on coastal ecosystems.

My first thoughts were over how to design a field research program that would put the hypothesis to a critical test. I knew there were other islands in the Aleutians besides Buldir that lacked foxes, either because they had not been introduced or because the introductions had failed. A proper pairing of these fox-free islands with those that were fox infested was the obvious way forward, and together with Vernon I could easily do that. But in addition to island selection, I would also need to determine what to measure and how to measure it, and here I would need help from others who knew more than I did about seabirds and terrestrial ecosystems. I approached Don Croll, an assistant professor at UC Santa Cruz with interests and experience in seabird biology and island conservation. Don liked the idea and suggested asking John Maron, a terrestrial ecologist from the University of Washington, to join us. John was also intrigued and keen to participate. So, including Vernon, who had traveled extensively through the Aleutians and knew the islands better than anyone, we had a first-rate team with all the talent and skills necessary to take on the project.

Our first challenge was deciding which islands to measure and how to measure them. Ideally, we wanted a mix of fox-infested and fox-free islands, spread across the Aleutian archipelago. This was important to ensure that whatever differences might exist between the two island groups were not

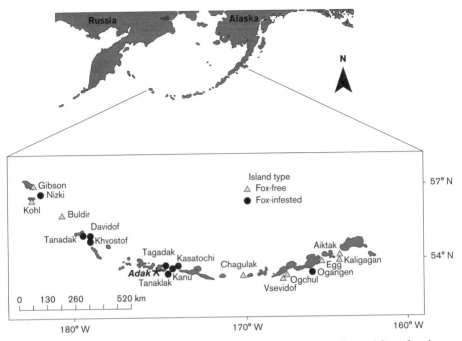

FIGURE 13.2. The Aleutian archipelago, showing the locations of fox-free and fox-infested islands that were surveyed in the fox–seabird study and Adak Island, where the nutrient-addition experiments were conducted.

confounded by something other than the fox effect. The size distributions of the two island groups had to be similar as well, because area-to-perimeter ratios increase with island size. This was important because nutrient input rate would be limited by island perimeter, whereas the deposition rate would be dictated by island area—and so, all else being equal, smaller islands should display a stronger fox–seabird–nutrient effect than larger islands. Fortunately, we were able to identify a suite of islands that met our needs (figure 13.2).

The next step was to figure out just what to measure and exactly where to measure it. Our hypothesis predicted greater seabird density on fox-free than on fox-infested islands. Fortunately, these data were already available from earlier monitoring by the Maritime Refuge in the Alaska seabird-colony catalogue (empirical data assembled by those who studied seabirds in Alaska and maintained by the USFWS). Our hypothesis further predicted differences in nutrient concentration between fox-infested and fox-free islands, so we would measure concentrations of nitrogen and phosphorus in the soils and plants. Methods for obtaining these measures were straightforward and widely used by plant and soil scientists, so here again, there was no challenge other than to gather the samples and have them analyzed.

Variation in nutrients predicted variation in species composition and biomass of terrestrial plants, which we would measure by simply identifying the species and clipping and weighing their aboveground biomass in sample quadrats. Lastly, our hypothesis predicted a proportionally greater source of marine nitrogen on fox-free islands, leading in turn to an expectation of enriched $\delta^{14}N$ in the soils, plants, and higher-trophic-level consumers. This expectation was based on the differing molecular weights of nitrogen's two stable isotopes, ^{13}N and ^{14}N (the latter being heavier than the former by the mass of one neutron), and a resulting proportional loss of the more common ^{13}N to the rarer ^{14}N from the breakdown and reconstruction of nitrogen-containing organic compounds as they move from prey to consumer upward through the food web. On average, an atom of marine-derived nitrogen goes through more trophic transfers than an atom of terrestrially derived nitrogen before mineralization. These processes and the resulting differences in $\delta^{14}N$ were well established prior to our study. We would thus measure $\delta^{14}N$ in soils, plants, and a range of consumers: slugs, flies, spiders, and passerine birds (via droppings beneath their perches).

How to measure the islands was a knotty issue. We couldn't simply go ashore and pick a spot that looked good, because even the most dull-witted critic would fairly see this as "cherry picking." We needed to array our sample locations over the various islands in a manner that would provide data that were as representative of each island as possible. In addition, we needed to design our sampling procedure to be conducted quickly and efficiently, because we would travel to and from the islands via ship, and ship time would be limited. We figured that we would have no more than a couple of days to sample each island. From a pilot sampling study, we knew about how much time would be needed to sample a single site on any given island; and considering the ship's size and the number of available berths, we knew that we would be able to deploy three or perhaps four sampling teams. Knowing all of this, we chose to superimpose a square array of points over the island, from which a random sample of 10 to 15 sites would be selected to actually sample. We would establish the exact geographic coordinates of each sample site beforehand and then use handheld GPS units to locate the sites.

Although we had good reason to believe that nutrients were intrinsically limiting to plant growth and production in the Aleutian Islands, this idea also needed to be rigorously demonstrated. We chose to test for nutrient limitation by adding a slow-release, nitrogen–phosphorus fertilizer on Adak—a large, fox-infested island—and contrast any changes in the plant community with those in unfertilized control plots. The plots would be fertilized in spring, before initial green-up, and then monitored in midsummer and in the two years following.

The final challenge in preparation for the study was to obtain funding. But the idea was a good one, and our proposal was selected for support by the National Science Foundation. We set out in 2001 with two technicians and a crew of student volunteers to begin the fieldwork. We would work on land during the days and steam between islands at night. Although I was excited by the upcoming adventure and the prospect of all we might learn, I was also daunted by the large number of islands to be visited in a relatively short time. In particular, I worried about weather and our ability to move people safely on and off the beach in rough seas. I also worried about the time it would take us to traverse the often steep and rugged terrain and whether we would, in fact, be able to reach and sample all the sites in the allotted time. And finally, I worried about habitat variation *(beta diversity)* and whether the resulting sampling error would swamp an otherwise important signal of variation in seabird abundance and nutrient input in our interisland contrast.

The National Science Foundation granted us authority to charter the *Tiglax* from the U.S. Fish and Wildlife Service because all their own vessels were then committed to the GLOBEC (Global Ocean Ecosystem Dynamics) program. This was a stroke of terrific luck, because the *Tiglax* crew knew the Aleutians and were highly skilled in landing and retrieving people and equipment with inflatable skiffs through the surf. In calm weather, skiff landings were safe and easy—but they could be an adventure when the sea was up.

Some of the islands were easy to sample, but others were more difficult. I came to almost dread our days on the fox-free islands, which were tough to hike across because of high vegetation. Fox-free islands also reeked of guano. Beyond that, steep islands were much harder to work on than the flatter ones. Chagulak—a precipitous, fox-free island with a single landing site and a nearly vertical scree slope from that point to the only piece of real estate that was flat enough for walking upright—was the worst of them all. But the islands were also magical in their wild beauty, each one being different enough from the last to turn every new day into an adventure (figure 13.3).

FIGURE 13.3. Eric M. Danner (right) and the author on the rim of Kasatochi Island's caldera. Kasatochi erupted several years after this photo was taken, covering the island in ash and extinguishing virtually all associated life.

We plodded along from island to island, and after three summers we had succeeded in sampling all the planned sites on all 18 islands, without significant mishap, except for one badly sprained ankle.

Not surprisingly, the 18 islands varied considerably from one another and across their individual landscapes. We expected this and had designed our study to capture the spatial variation that inevitably occurs in all natural ecosystems. We knew that this variation would be substantial and wondered whether it would render any effects of foxes and seabirds invisible. It did not. The data confirmed our hypothesis and what we already knew from the seabird-colony catalogue—and knew or suspected from casual observation during our visits to the various islands. Seabird densities were nearly two orders of magnitude greater on the fox-free islands than on the fox-infested islands. Although the total areas of the two groups of islands were similar, the fox-infested islands supported an estimated 15,000 breeding seabirds whereas the fox-free islands supported more than half a million.

Soil nutrient concentrations (as measured by total extractable phosphorus and nitrogen) were higher on the fox-free islands. But the signal was clearer for phosphorus than it was for nitrogen. Extractable phosphorus concentrations were about threefold higher on the fox-free islands, whereas the difference in soil nitrogen concentration was only about 25 percent. We can't explain the difference between soil phosphorus and nitrogen with certainty. It may be that the mobility of soil nitrogen is greater than that of soil phosphorus because of differences in leaching or plant uptake rates. Or it may be that nitrogen is returned to soil through another pathway (e.g., from the atmosphere by nitrogen-fixing plants). We suspected differences in leaching or uptake, but in truth we still don't know.

We focused exclusively on nitrogen and not on phosphorus in plant tissues because nitrogen was considered the more critically limiting of the two nutrients to terrestrial plant growth, and leaf nitrogen concentrations had been shown in many previous studies to vary with nitrogen availability. We measured total leaf nitrogen concentration in selected grass and forb species and found significantly elevated levels on the fox-free islands.

Different nutrient availabilities had clear effects on plant communities between fox-infested and fox-free islands, as reflected in species composition and total plant biomass. Although grasses were the numerically dominant plant type on fox-infested islands, these grasses were often spindly-looking and were matched in abundance by a diverse assemblage of low-growing shrubs, forbs, mosses, lichens, and various other, less common plant types. Fox-free islands, by contrast, were heavily grass-dominated, with shrubs, forbs, mosses, and the various other types comprising less than 20 percent of the overall plant community. Total plant biomass was about 2.5 to 3.0 times greater on the fox-free islands. Differences in the plant communities between fox-infested and fox-free islands were striking to the glance, the former being a lightly hued maritime tundra and the latter a richly green maritime grassland.

The fertilization experiments on Adak confirmed that differences in the plant communities between fox-infested and fox-free islands indeed resulted from differences in nutrient limitation and availability. We first applied fertilizer to the treatment plots in May 2002. Increased greening and plant growth were obvious when we returned to Adak in July. By 2004, the plant communities in the nutrient-addition plots were grass dominated and had increased by severalfold in total plant biomass (figure 13.4). Over the same period, the unmanipulated control plots remained unchanged. These effects were strong enough to be visible in satellite images several years later. I continued to visit the plots over the years, whenever I was on Adak. The fertilized plots were distinctly visible through 2010, but I was no longer able to detect them when I last looked in early August 2014.

FIGURE 13.4. Fertilization experiment on Adak Island. The visually distinct square plots are those to which fertilizer was added.

We had little doubt that the elevated nutrients on fox-free islands were derived from the sea. But we would have been remiss in not testing that hypothesis more explicitly, because nitrogen source was easy to determine through stable isotope analysis. If the nutrients on fox-free islands were indeed being derived disproportionately from the sea, we ought to see a consistent enrichment in $\delta^{14}N$ throughout the ecosystem—from soils, to plants, to the various terrestrial consumers that were ultimately nourished from terrestrial plants' primary production.

I was leery of what the isotope measurements would show, because the field survey data already told a compelling story; I wondered how we could reconcile these data with the isotope data if the isotope measurements didn't fall into line with what we expected to see (and I had been duped before by seemingly simple and straightforward expectations that hadn't worked out). But the only real surprise in the isotope data was how strongly and consistently they supported everything else. For soils, plants, and all the various consumers we contrasted between fox-infested and fox-free islands, $\delta^{14}N$ was enriched by about 3–6 per mill (‰), which indicates a difference of one or two trophic levels between sources—the general difference other scientists had measured before, and have consistently measured since, between land- and marine-derived nitrogen in terrestrial ecosystems.

The story to emerge from this body of research was both empirically well substantiated and, to my mind, about as cool as anything I had ever been involved with. The introduction of foxes to the Aleutian Islands in the early 1900s devastated the seabirds, reducing their overall number by a hundred-fold. The reduced seabird numbers reduced nutrient input from an enriched ocean to an intrinsically nutrient-impoverished land, thereby driving a shift in the terrestrial plant community from a productive maritime grassland to a comparatively unproductive maritime tundra (Croll et al., 2005).

My colleagues and I were excited by these findings for two main reasons. First, they established a link between predators and plants that occurred via a fundamentally different pathway from the traditionally understood trophic cascades that had grown out of the "green world hypothesis" (Hairston et al., 1960). Classical trophic cascades worked exclusively through top-down forcing—the limiting influences of consumers on their prey and the linking together of these consumer–prey interactions across successively lower trophic levels. Our fox–seabird work linked apex predators with plants through a chain of interconnected consumer–prey interactions, and did so via a key element of bottom-up forcing (effects of nutrients on plants) and a linkage across adjacent ecosystems (the transport of nutrients from sea to land via seabirds). A second notable feature of our findings was the scale of the demonstrated effect. Fox introductions influenced island plant communities in a broadly consistent manner over a stretch of at least a thousand miles across the Aleutian archipelago. While such a large spatial scale of impact isn't surprising, it was unprecedented in experimental studies of species interactions.

For me, the reason for wonder and excitement extended beyond these patterns, the biological players, and the ecological processes because the work established yet another dimension to the manner in which the sea otter's influence spread through its surrounding ecosystem. I had begun in the early 1970s by imagining sea otters as inconsequential players in the organization and control of their surrounding ecosystem. Discovery of the otter–urchin–kelp trophic cascade changed that perspective fundamentally but left me still with a view of ecological process that did not extend beyond the coastal kelp forest ecosystem. The discovery of interactions between sea otters and lumpsuckers, and shortly thereafter between sea otters and killer whales, expanded my perspective of process to include important, biologically driven links between the nearshore coastal ocean and the open sea. All along, the land seemed isolated from the sea otter's sphere of ecological influence, perhaps because the physical media (water versus rock and soil) and the things that lived there were so different. But in fact, the sea otter was a key player in the fox–seabird story and the transition of terrestrial ecosystems from maritime grasslands to maritime tundra—simply because, if the sea otters had not been overhunted, foxes probably would not have been introduced to Alaska islands.

This journey of learning has taught me that the ecological connections from predators to their prey spread through surrounding nature in ways that are amazingly complex and profoundly important to how the world looks and operates.

FOURTEEN

A Global Perspective

FIELD ECOLOGISTS FOLLOW DIFFERENT PATHS in their efforts to understand the workings of nature. Some move from system to system, by design or as their interests change. Others spend all or most of their lives trying to understand a single species or ecosystem. I chose the latter pathway because the science held my interest, because I loved the Aleutian Islands, and because I could. Each new discovery raised exciting new questions. Serendipity also had a lot to do with it, as the unexpected collapse of sea otter populations in southwest Alaska defined much of my interest and research program over the past two decades.

My intellectual and emotional love affairs with the Aleutian Islands grew over the years with increased familiarity and understanding. When I gaze upon or think about the Aleutians today, I see not only the patterns whereby species are spread across the landscapes and seascapes, but what in the beginning had been a largely invisible infrastructure of interactions among these species with one another and their physical environment. I have grown to understand why kelp forests abound in some areas and are absent in others; why sea otter populations recovered and then declined over the past century; and why some islands are festooned with grasses while others are covered with maritime tundra. Body and mind willing, I might have continued down this path to the end of my professional life. But I was drawn to think more broadly about how nature works.

Part of that draw was simple curiosity. I couldn't help but wonder if the same sorts of interactions that linked sea otters to kelp forests in the North Pacific occurred elsewhere in nature. But another part of the draw was really more of a push, motivated largely by the fallout of the whale wars (see chapter 12). I had grown weary of my marine colleagues' seemingly pedantic ecological mind-sets and their often cynical or even hostile reactions to my view of ecological process. I could have continued in my own small world, but I knew that many of the scientists, managers, and policy makers with whom I was closely associated would never take my view of nature seriously. Even my close friends from the marine mammal world were skeptical. I needed to somehow move forward in a way that was responsive to this push-and-pull. I perceived this need, but the way forward was not clear.

My new path began to open up and assume shape with a message that fell out of the sky—in the form of an email from John Terborgh, whom I knew mostly by his reputation as one of the world's foremost tropical ecologists. He had come to believe that predation and top-down forcing were fundamentally important to the ecology and evolution of the *Neotropical* forest systems he had spent a lifetime studying—just as I had found in the kelp forests. And like me, he was frustrated by the many tropical ecologists who resisted his views. I never asked why he chose to reach out to me. Part of it was probably his awareness of our common view of process in nature and our common struggle for acceptance of our ideas. Whatever his reasons, he wrote with a simple idea that struck me as powerful and exciting. John's proposal was to identify and assemble a group of senior ecologists who knew or studied different systems around the world for the purpose of assessing the roles of predation and top-down forcing across global ecosystems. I immediately agreed to join him in the endeavor. This was the way forward that I had been searching for.

Our initial tasks were to settle on invitees, decide where and when to hold the meeting, and figure out how to fund it. John began the search for a venue by asking the White Oak Foundation to host our meeting at their facility near Jacksonville, Florida. They agreed. The next step was to assemble a list of potential invitees. We did this by first listing what we saw as the major global ecosystem types and then identifying scientists who knew the workings of those various systems. Our goal was to fairly represent the range of planetary habitats, from the tropics to high latitudes and from land and fresh water to the sea. We also included several people with more conceptual or theoretical interests and backgrounds. White Oak limited us to a maximum of 20 participants, which forced us to choose carefully and exclude some people we would have liked to include. There was no clear choice in many cases, so we picked people we knew and who we trusted would contribute to the workshop and its proceedings in a positive and productive way. Nearly all

Participants in the White Oak Meeting

Participant	Area of Expertise
Joe Berger	Large ungulate behavioral ecology
William J. Bond	Plant ecology
Justin Brashares	African savannas
Stephen R. Carpenter	Lakes and freshwater ecosystems
Timothy E. Essington	Oceanic systems
James A. Estes	Higher-latitude coastal marine ecosystems
Robert D. Holt	Food web theory
Jeremy B. C. Jackson	Tropical reefs
Robert J. Marquis	Terrestrial plant–herbivore systems
Lauri Oksanen	Terrestrial arctic ecosystems
Robert T. Paine	Rocky intertidal ecosystems
Ellen K. Pikitch	Sharks and marine ecosystems
William J. Ripple	Temperate–boreal terrestrial ecosystems
Marten Scheffer	Stability theory; lake ecosystems
Thomas W. Schoener	Tropical islands; food webs and foraging theory
Jonathan B. Shurin	Trophic cascades across ecosystems
Anthony R. E. Sinclair	African savannas; boreal–subarctic ecosystems
Michael E. Soulé	Chaparral habitats; conservation biology
John Terborgh	Tropical forests
David A. Wardle	Soil and plant ecology

of the initial invitees enthusiastically agreed to join us (see table). Island Press expressed interest in publishing the proceedings.

White Oak would provide food, lodging, and a meeting room, but we needed additional funding for travel to and from Jacksonville. The securing of these funds was our final obstacle. After several inquiries and follow-up discussions, the Pew Charitable Trusts and Defenders of Wildlife generously provided the required support.

Our requests of the participants were straightforward and simple—give an overview presentation describing what was known or suspected about the major controlling processes in the systems they had studied over the course of their lives, and attend the workshop with an associated manuscript in hand.

We met in February 2007. As one speaker after the next described the overall workings of the systems they had studied and knew well, we all began to see evidence of what John and I, at least, had suspected from the outset—strong effects of apex consumers on the structure and function of ecosystems all over the world. Trophic cascades were a recurrent theme in these accounts, as were their wider-ranging influences on other species and ecological processes. Study methods, detailed results, and the strength of evidence varied from system to system. But the evidentiary deficiencies and resulting uncertainties that plagued an understanding of the workings of any particular ecosystem were greatly diminished by a commonality of pattern and process across ecosystems.

We began the workshop knowing that at least some ecosystems were strongly influenced by predators and top-down forcing processes but not knowing how widespread or important these effects were across a representative suite of global ecosystems. Our intent in the beginning was to report the results in the workshop's proceedings, whatever they might be. By the end, we had come to a strong and exciting sense of commonality in ecological process. Predators and top-down forcing were important across all or most of nature, and the systematic loss of these species and processes had changed and were continuing to change the world we live in. The edited volume of the proceedings would make these points, but many participants felt that a book alone would not fully convey our message to science and society. In the workshop's final hours, we decided to write a shorter, overview account that might be of sufficiently broad interest for a wide-reaching journal like *Science* or *Nature*. The task of leading that effort eventually fell on me.

Although we all had listened to the same presentations and had read the same summary manuscripts, each of us processed this information from different reference points. Mine was the sea otter–kelp forest system, from which I saw predators limiting herbivores and, in turn, facilitating autotrophs. I further saw these effects occurring as abrupt phase shifts characterized by hysteresis as otter populations waxed and waned. I saw the effects of trophic cascades spreading to other species and ecological processes through the creation or destruction of biogenic habitat, altered production, and influences on the physical environment through the attenuation of waves and currents and increased sequestration of carbon dioxide. I saw consumer–prey systems in an evolutionary context, through the decoupling effect of predators on the coevolution of defense and resistance in plants and their herbivores, and in the resulting influences of plant quality on the radiations of other plant consumers. Finally, I saw the kelp forest ecosystem as not only held together through local interactions but connected with species and events in other places and at other times. This was my view of nature, and my intent going forward was to explore how what others had learned and thought about their respective ecosystems might map onto that worldview. I worried some about the old adage of finding what one looks for. But I knew that my colleagues would be looking carefully and critically over my shoulder, so I didn't worry too much.

TOP-DOWN FORCING AND TROPHIC CASCADES

The White Oak Workshop's principal objective was to explore the occurrence, and where possible the relative importance, of top-down forcing and trophic cascades across global ecosystems. I learned that there is nothing unique or unusual about the sea otter–kelp forest trophic cascade, except perhaps in the species themselves. Top-down forcing and trophic cascades occur everywhere in nature. I've gathered what I see as the high points and present these in a narrative overview for interested readers at the end of this chapter (see box 2).

PHASE SHIFTS AND HYSTERESIS

As sea otter densities increased or declined with growth or collapse of populations at various locations across the Pacific rim, their effect on sea urchin and kelp abundance occurred not as a gradual change, but as a distinct phase shift. Particular areas on the seafloor, and by extension entire reefs, entire islands, and even entire regions, are occupied by either kelps or urchins, depending largely on the status of associated sea otter populations. These phase shifts occur so rapidly that they are seldom observed directly. Kelp forest phase shifts are characterized by hysteresis, or differing trajectories of change with otter population density, depending on whether the otters were increasing or declining. Moreover, the otter population densities at which the phase shifts occur, and thus the patterns of hysteresis, differ markedly in different parts of the North Pacific Ocean (see chapter 6). To what extent do nature's diverse trophic cascades occur as phase shifts? And if phase shifts are common, how many are characterized by hysteresis?

In most cases we don't yet know the answer to those questions, but for several we do. Thanks to creative work by Marten Scheffer and colleagues, temperate lakes provide the most thorough understanding of phase shifts and hysteresis (Scheffer et al., 2001). The majority of shallow, temperate lakes around the world tend to be either clear (owing to a dearth of phytoplankton) or opaque green (owing to an abundance of phytoplankton). Phytoplankton abundance is controlled, in turn, by two main forces—from the bottom up by nutrient concentrations and from the top down by zooplankton grazing. Scheffer's analyses are based mostly on changing nutrient levels and their influence on lake turbidity. Pristine lakes are *oligotrophic,* which means that phytoplankton are comparatively rare, lake water is clear, and sunlight is able to penetrate to the lake floor, where it fuels the growth of a benthic (bottom-dwelling) plant community. The benthic plants help maintain lake clarity through three processes: competition with phytoplankton for limiting nutrients; protection of zooplankton from fish predators, thereby increasing consumption rates of phytoplankton by zooplankton; and dampening wave action on the lake floor, thereby reducing sediment suspension and associated turbidity. Turbidity increases with nutrification, but the functional relationship between turbidity and nutrient level is lower for lakes with benthic plant assemblages than for lakes that lack this assemblage, as explained above. Lastly, there is a critical turbidity level above which light penetration to the lake floor is insufficient to maintain the benthic plant assemblage. As a result, lakes tend to either retain or lack a benthic plant assemblage, and the shift between these states is characterized by hysteresis (figure 14.1). Lakes with intact benthic plant assemblages tend to remain relatively clear, even at elevated nutrient concentrations. But once nutrient levels have increased turbidity to the point that benthic plants can no longer survive, the system quickly flips to a benthic plant–free state, and nutrients must be reduced to levels well below those that triggered the change before this phase shift is reversed.

To my knowledge, the wolf–elk–willow assemblage is the only other predator-driven system for which there is reasonable evidence of hysteresis. That evidence comes from experimental and long-term observational studies of riparian habitats following the repatriation of wolves to Yellowstone National Park (Marshall et al., 2013). After the loss of wolves from Yellowstone early in the twentieth century, elk proliferated and eventually reduced willow abundance in riparian habitats through intense overgrazing. This, in turn, caused the water table to drop, in part because of an associated loss of beavers caused by overtrapping and the absence of streamside woody vegetation needed for constructing their dams. Marshall and colleagues showed that the

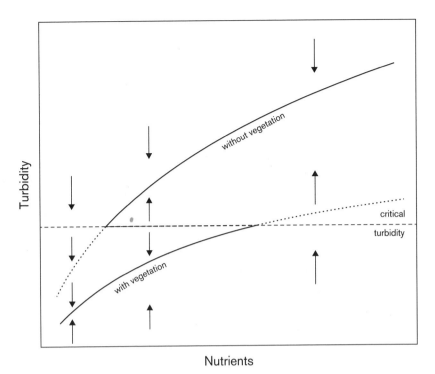

FIGURE 14.1. Functional relationship between nutrient loading and water turbidity in lakes. For explanations of the mechanisms, see chapter narrative and Scheffer et al. (2001).

water table's drop slowed willow recovery, even in areas protected from elk grazing. The pattern of willow recovery following wolf reestablishment and reduced elk grazing is probably quite different from the pattern of increasing elk abundance and willow loss that followed the loss of wolves from Yellowstone.

INDIRECT EFFECTS OF TROPHIC CASCADES

Effects of the otter–urchin–kelp trophic cascade extend widely across coastal ecosystems to influence numerous other species and ecological processes (see figure 8.1). For the most part, these various and sundry ecological influences spin off the sea otter's enhancing effect on the distribution and abundance of kelp, which in turn provides habitat for other species, increases ecosystem productivity, modulates the force and velocity of ocean waves and currents, and intercedes in the cycling of chemical elements and compounds through the ecosystem. Although kelp forests themselves are the primary node through which the effects of sea otter activity spread to other species, that's not always the case. In some instances the sea otter's indirect ecological influences spin off the altered abundance of higher-trophic-level species, as ensues from their dietary effect on glaucous-winged gulls. In some instances the sea otters' indirect ecological influences occur by way of altogether different food-web pathways, as happened when their depletion of sea stars around Attu enhanced survival rates of mussels and barnacles. Trophic cascades occur widely across global ecosystems. How pervasive are the indirect effects of these various trophic cascades on other species and ecological processes?

We don't yet know the answer to that question, even for the sea otter–kelp forest system. But there are intriguing hints of diverse effects in many systems, and reasonably strong evidence from a few of these. The most thoroughly studied companion cases are in New World temperate and boreal forests, spun off from the inhibitory effects of wolves on elk and coyotes. Wolf predation on elk facilitated plant growth and created a carrion resource for scavengers. Riparian-zone plant losses that followed wolf eradication altered stream courses and lowered water tables (Beschta and Ripple, 2006) on one hand; on the other hand, these losses either reduced populations or eliminated species that depended in some way on riparian vegetation (such as beavers and nesting passerine birds; Berger et al., 2001). Intense elk grazing in upland habitats reduced berry production, in turn reducing berry consumption by grizzly bears (Ripple et al., 2014). Wolf-created elk carcasses provided food resources for scavengers, such as bears and ravens (Wilmers and Getz, 2005).

The inhibitory effect of wolves on coyotes through direct killing and behavioral avoidance enhanced fawn survival in pronghorn (Berger et al., 2008) and may have enhanced lynx populations through reduced competition with coyotes for snowshoe hare, the lynx's most important prey species (Ripple et al., 2011). The spread of coyotes into the midwestern and northeastern United States has been followed by widespread declines in red fox, purportedly resulting in eruptions of the fox's small mammal prey and the follow-on rise in the ticks that depend on these small mammals for a blood meal during the earlier states of their life cycle (Levi et al., 2012). The epidemic rise of *Lyme disease* is strongly correlated with the rise of coyotes and the decline of foxes, and thus may well be linked to the loss of wolves.

Other effects are known or suspected among other species of predators or large consumers in other ecosystems. The decline and subsequent recovery of large ungulates in East Africa that resulted from the spread, and ensuing control, of the viral disease rinderpest influenced the frequency and intensity of wildfire, which followed directly from variation in fuel load caused by variation in the intensity of ungulate grazing pressure (Holdo et al., 2009). Increases in sea urchin population densities that followed depletion of predatory lobsters in Southern California caused increases in the epidemic frequency of urchin wasting disease (Lafferty, 2004). These increases likely resulted from elevated rates of disease transmission among individual urchins because of their closer proximity to one another in predator-free environments. Midwestern lakes with bass absorbed less carbon dioxide from the atmosphere than the same lakes without bass, owing to the effect of the bass-induced trophic cascade on phytoplankton abundance and the uptake of carbon dioxide by photosynthesizing phytoplankton (Schindler et al., 2008). Spawning salmon in rivers and streams of the North American Pacific Northwest have higher downstream sediment loads than streams and rivers that lack salmon, owing to the sediment-clearing effect of salmon on their upstream spawning beds (Moore et al., 2007). The extinction of native birds in Hawaii has increased *invasibility* of these systems to nonindigenous spiders (Gruner, 2005). And, finally, the loss of coyotes from chaparral habitat fragments in Southern California has been accompanied by sharp declines in the diversity of small vertebrate species, resulting in large measure from the rise of smaller predatory mammals, including domestic cats, which occurred in response to the loss of coyotes (Crooks and Soulé, 1999).

Although the evidence is scattered and of varying quality, the conclusion from these various examples was clear to me. The sea otter–kelp forest system is unique only in the details of the associated species' natural histories and the physical environment in which they occur. Diverse and important indirect effects of trophic cascades occur broadly across our planet's ecosystems.

EVOLUTION

Direct evolutionary influences of consumers on their victims and the reciprocal coevolutionary consequences of consumer–victim interactions are known from numerous species in a diversity of ecosystems. Work by Nick Davies and colleagues on nest parasitism by European cuckoos provides one of the clearer and better-understood examples (Brooke and Davies, 1988). Cuckoos lay their eggs in the nests of various passerine host species, which amazingly (to the human eye) are unable to recognize cuckoo eggs and cuckoo chicks as being different from their own. Cuckoo chicks not only compete with their nest mates for food delivered by the mother to nourish them—they kill the mother's own young by ejecting them from the nest. Over time, in response to this fitness cost, the hosts evolve an increased ability to recognize the parasite's egg and thus increasingly eject the egg, thereby imposing a fitness cost on cuckoos. The result is an evolutionary "arms race" in which individual cuckoo lineages specialize in particular host species and evolve eggs with increasingly similar color patterns to those of their hosts. The process is virtually identical to the coevolution of defense and resistance in plant–herbivore systems, which are also widely known in nature. However, coevolutionary consequences of the indirect effects of trophic cascades are almost entirely unknown in nature (Estes et al., 2013). To my knowledge, the work of Peter Steinberg and myself on the coevolutionary consequences of trophic cascades (see chapter 9) provides the only example of this phenomenon. Coevolutionary consequences of the indirect effects of predators may be rare in nature. Conversely, they may be common but poorly known because people haven't looked for them.

INTER-ECOSYSTEM COUPLING

A final major dimension to the work my colleagues and I have done on sea otters and kelp forests involves the linkages between coastal ecosystems and the oceanic realm. We saw such linkages in food subsidies to sea otters by episodic inshore spawning migrations of smooth lumpsuckers, and in the depredation of sea otters by killer whales (see chapter 10). Similar linkages across systems are known for several other species and ecosystems. Merav Ben-David has shown that North American river otters enhance plant production around their forest latrine sites with marine-derived nutrients (Ben-David et al., 2008), much as Pacific salmon deliver marine nutrients to the forests surrounding their spawning streams (Naiman et al., 2002). Tiffany Knight and colleagues experimentally demonstrated a strong link between predatory fish in central Florida lakes and plant reproductive success on nearby lakeshores (Knight et al., 2005), in which case the fish prey on dragonfly larvae, thereby reducing adult dragonflies around the ponds and enhancing the dragonfly's insect prey, which in turn pollinate lakeside plants. Doug McCauley and Hilary Young, working at Palmyra Island in the remote tropical Central Pacific Ocean, have shown that seabirds enhance terrestrial production by vectoring guano from sea to land; that this effect is stronger in the native plant community than in exotic coconut palm forests; that nutrient return from land to the highly oligotrophic nearshore waters is similarly higher adjacent to native plant forests than adjacent to the exotic coconut palm forests; that because of this nutrient return, coastal marine production and, thus, zooplankton abundance are also greater in coastal waters adjacent to the native forests than in those adjacent to the coconut palm forests; and, hence, that planktivorous manta rays are more common in the coastal waters adjacent to native forest communities (McCauley et al., 2012). I could recount many additional examples. But these are among my favorites and are sufficient to demonstrate that consumer-induced linkages occur in diverse ecosystems.

In the preceding pages and in box 2, I've summarized enough evidence to demonstrate that top-down forcing and trophic cascades are a recurrent process in nature and that the indirect effects of trophic cascades spread widely through ecosystems to numerous other species and ecological processes; that these various ecological processes have important evolutionary consequences, at least sometimes; and that predators and their ecological effects sometimes link, in important ways, what are usually thought of as distinct ecosystems.

None of this is to say or imply that other ecological processes are either unimportant or secondary to top-down forcing and trophic cascades in dictating the distribution and abundance of species across global landscapes and seascapes. Every reasonable ecologist knows that numerous factors influence these patterns of life on Earth. Nonetheless, consumer–prey interactions are ubiquitous across living nature, and the associated influences of top-down forcing and trophic cascades are often evident through all of nature's complexity. Nor do I argue that top-down forcing and trophic cascades occur everywhere and at all times in nature. They may, but the evidence is insufficient (as it always will be) to make such a claim. But I think it is fair to say this: The sea otter's story is not unique or even unusual. Top-down forcing and trophic cascades occur widely in nature and have been found everywhere people have carefully looked for them.

I've organized the following information in accordance with our planet's three major habitats: oceans, fresh water, and land. This account was written in collaboration with Justin Brashares (terrestrial ecosystems) and Mary Power (freshwater ecosystems) (Estes et al. 2013).

Ocean Ecosystems

The world's oceans can be divided into four broad regions or habitat types: intertidal zones, shallow temperate seas, shallow tropical seas, and the vast oceanic realm. Although this grouping necessarily excludes an immense amount of important detail, most ocean ecosystems can be assigned to one of the four categories.

Intertidal zones are the empirical birthplace of our understanding of the ecological consequences of predation, top-down forcing, and trophic cascades. That understanding grew from pioneering experimental studies by people like Joe Connell and Bob Paine, who demonstrated the importance of predatory snails and sea stars by removing these species and documenting subsequent changes on both their prey and the associated community of organisms. Paine's classic studies of rocky intertidal communities on the outer coast of Washington chronicled a loss of diversity in areas from which ochre stars *(Pisaster ochraceus)* had been removed, owing to the fact that *Pisaster*'s preferred prey, the sea mussel *(Mytilus californianus),* was also a competitive dominant—which, in the absence of *Pisaster* predation, excluded other species of intertidal invertebrates and algae (Paine, 1974). A plethora of broadly similar findings has since appeared in the scientific literature from studies of other predators in rocky intertidal systems around the world. Phil Hockey's work in South Africa, which demonstrated a trophic cascade from predatory black oystercatchers to herbivorous limpets to intertidal algae, is one of the many cases in point (Hockey and Branch, 1984). Similar findings have emerged from Brian Silliman and colleagues' work in salt-marsh ecosystems of the southeastern United States, where previously unexplained die-offs of the dominant plant, saltmarsh cordgrass, were shown to have resulted from declines in predatory crabs and the resulting increase of herbivorous snails (Silliman and Bertness, 2002).

The influences of predation (by invertebrates, sharks, and other large fishes) and herbivory (mostly by fishes and sea urchins) on tropical coral reefs and seagrass meadows are well and widely established (Sandin et al., 2010). "Grazing halos," bands of bare sand that separate Caribbean patch reefs from surrounding turtle grass meadows, are caused by herbivorous fishes and sea urchins that shelter in the reefs by day and venture forth to feed by night (Randall, 1965; Ogden et al., 1973). Observational (Williams and Polunin, 2001; Mumby et al., 2006; Newman et al., 2006) and experimental (Smith et al., 2001; Thacker et al., 2001; McClanahan et al., 2003; Burkepile and Hay, 2006; Hughes et al., 2007) studies have demonstrated a primary role of herbivores in limiting autotrophs on coral reefs in the Atlantic, Pacific, and Indian oceans. Cryptic coloration (Hixon, 1991; McFarland, 1991), schooling (Alexander, 1974; Magurran, 1990; Sandin and Pacala, 2005), and refuging (Ogden et al., 1973; Hixon and Beets, 1993; Friedlander and Parrish, 1998) are widespread evolutionary responses that reduce the risk of predation for reef dwellers. Predator-prey and herbivore-autotroph interactions are linked as trophic cascades in the Caribbean (Hughes, 1994), Gulf of Mexico (Heck et al., 2000), western tropical Pacific (Dulvy et al., 2004), and Indian Ocean (O'Leary and McClanahan, 2010). These processes are particularly evident from Doug Rasher and colleagues' recent work on Fijian reefs where tribally maintained marine reserves are dominated by corals while adjacent, heavily fished areas are overgrown with algae (Rasher and Hay, 2010).

Consumer effects are especially well studied in temperate to boreal kelp forest systems (Estes et al., 2010a). The collapse of overfished cod stocks led to population irruptions by herbivorous sea urchins and kelp deforestation in the Gulf of Maine (and probably elsewhere) in the western North Atlantic (Steneck et al., 2004). In the absence of sea otters, Southern California kelp forests are now maintained by various consumers, including benthic predatory fishes (Nelson and Vance, 1979; Cowen, 1983), lobsters (Tegner and Dayton, 2000; Lafferty, 2004), and zooplanktivorous fishes (Davenport and Anderson, 2007).

Similar patterns and processes occur in higher-latitude coastal marine systems of the Southern Hemisphere (Steneck et al., 2002). Predation by recovering populations of fishes and lobsters has led to sea urchin reductions and kelp recovery in New Zealand marine reserves (Babcock et al., 1999). Kelp forests have collapsed in parts of Tasmania because of the interactive influences of overfishing of predatory lobsters and the southward range expansion of warm-temperate sea urchins (Ling et al., 2009). In South Africa, the loss of predatory lobsters (apparently caused by an anoxic event) led to increases in predatory whelks (previously preyed on by lobsters) and a predator–prey role reversal, in which predation by groups of these whelks overwhelmed colonizing lobsters, preventing their establishment (Barkai and McQuaid, 1988).

Evidence of trophic cascades involving pelagic predators in oceanic ecosystems is sparser but increasing (Baum and Worm, 2009; Essington, 2010). In nearly all cases, the key data come from time-series measurements associated with the natural fluctuations of predators or their depletion in fisheries. The loss or reduction of great sharks from eastern U.S. coastal oceans and estuaries triggered a trophic cascade in which the great sharks' prey (smaller sharks, skates, and rays) have since irrupted, thus sharply reducing populations of the infaunal bivalve mollusks that are preyed upon by these smaller elasmobranches, in turn causing the collapse and closure of various clam fisheries (Myers et al., 2007). Declines of filter-feeding bivalves may have caused phytoplankton to increase, thus reducing water clarity and quality (Jackson et al., 2001; Kirby, 2004). Reductions of finfish by fisheries in the Black Sea has led to an increase in small planktivorous fishes, a reduction in zooplankton, an increase in phytoplankton, and, ultimately, a regime shift to a system dominated by gelatinous plankton (Daskalov et al., 2007). Cod declines from overfishing across the North Atlantic Ocean have been closely followed by increases in their prey—shrimp (Worm and Myers, 2003), crabs (Frank et al., 2005), and zooplanktivorous fishes—which in turn led to reduced zooplankton abundance followed by increasing phytoplankton and oceanic chlorophyll concentrations (Frank et al., 2005, 2011; Casini et al., 2008, 2009).

The extirpation of great whales from the world's oceans has diminished their effects as consumers of krill, squid, and forage fish; as prey to apex predators such as giant sharks and killer whales; as carcasses delivering high concentrations of lipids and other nutrients to the seafloor in an otherwise nutrient-impoverished deep sea (Estes et al., 2006); and as vectors of various nutrients from deep waters to the *euphotic zone* (Nicol et al., 2010; Roman and McCarthy, 2010; Roman et al., 2014). Although still largely unrecognized by cetacean biologists and oceanographers, the ecological consequences of these changes are striking and diverse. Prior to industrial whaling, the great whales co-opted an estimated 60% or more of North Pacific Ocean's *net primary production* (Croll et al., 2006). The reduction of great whales in the Southern Ocean caused or contributed to a dietary switch in Adélie penguins from fish to krill (Emslie and

Patterson, 2007); and whale reductions in the North Pacific appear to have caused their foremost predators—killer whales—to expand their diet to include seals, sea lions, and sea otters, thereby driving populations of these prey to collapse (see chapter 10).

Freshwater Ecosystems

Freshwater ecosystems—flowing rivers and streams, and still lakes and ponds—have been important arenas for understanding food-web dynamics because of the relatively short generation times of lower-trophic-level species and their clearly circumscribed physical borders, which make them well suited for comparative and experimental analyses. Impacts of apex consumers are known in fresh water from the lowland tropics to high elevations and high latitudes. Understanding of these impacts comes both from purposeful experimental manipulations, in which naturally occurring consumers have been added or removed, and from time series that follow the introductions of exotic species.

Standing freshwater ecosystems (lakes and ponds) provide some of the most convincing and well-known evidence for trophic cascades. The earliest reports in the scientific literature come from floodplain ponds in what is now the Czech Republic, where Hrba ek and colleagues (Hrba ek et al., 1961) documented strong responses in phytoplankton, zooplankton, water chemistry, and ecosystem metabolism to changes in fish species. While the classic whole-lake experimental studies by Steve Carpenter, Jim Kitchell, and colleagues are a well-known textbook example, cascading effects of apex predators (usually fish) have been observed and experimentally induced in ponds and lakes throughout Europe and North America (Brooks and Dodson, 1965; Carpenter et al., 1985, 1987; Persson et al., 1993). These trophic cascades progress through planktivorous fish to grazing zooplankton to phytoplankton, the principal autotrophs in lake water columns. When *piscivorous* fish (commonly the lake's apex predators) eat zooplanktivorous fish, grazing zooplankters are released from predation and can clear lake water columns of phytoplankton in weeks or months. Without piscivores, zooplanktivorous fish (often minnows in temperate lakes) suppress these grazers, and phytoplankton builds up to turn lakes green. Such effects are evident and striking to the glance.

Trophic cascades also occur in rivers and streams where apex predators either release (Power et al., 1985; Flecker and Townsend,

1994; McIntosh and Townsend, 1996) or suppress (Power, 1990; Wootton and Power, 1993) stream algal biomass, depending on whether food chains have odd or even numbers of functional trophic levels (Fretwell, 1987). As with lakes, top-down impacts of predators in rivers vary across space and time. In the Eel River of Northern California, juvenile steelhead *(Oncorhynchus mykiss)* indirectly enhance algal biomass by consuming herbivorous insects, but only under normal, winter flood–summer drought hydrologic regimes. Winter floods sweep the streambed free of macroinvertebrates and algae, but fast-growing edible grazers recover quickly during the following summer, when they are eaten by various predatory fish. Drought years lack this winter flood scour, during which periods the longer lived and predator-resistant armored caddisflies, which are also herbivores, increase in numbers, thus breaking the predatory-fish-induced trophic cascade and causing algal biomass to decline (Wootton and Power, 1993; Power et al., 2008).

Trophic cascades are known in other freshwater ecosystems around the world from the introduction or removal of exotic consumers. Beavers *(Castor canadensis),* which maintain riparian water tables and wetland habitats, in turn enhance fish, riparian vegetation, and associated wildlife across many parts of North America. These functions have been seen and documented with the loss and subsequent recovery of beavers from the fur trade. The introduction of beavers into southern Chile in the mid-1990s had a very different effect, damaging a native flora that had not coevolved to withstand beaver impacts and facilitating the spread of invasive plants into formerly pristine ecosystems of southern Chile (Anderson et al., 2006, 2009). Similarly, introduced nutria *(Myocastor coypus)* from South America are spreading across the U.S. Gulf Coast, where they are damaging marsh vegetation, including island-colonizing trees essential for land building (Fuller et al., 1985; Carter et al., 1999). Invasions of sea lamprey *(Petromyzon marinus)* into North American Great Lakes after construction of the St. Lawrence Seaway in the late 1800s subsequently devastated lake trout and associated Great Lakes fisheries (J. Kitchell, cited in Burton, 2010). Nile perch *(Lates niloticus)* were introduced into Africa's Lake Victoria, where they have exterminated hundreds of endemic haplochromine cichlid species (Kaufman, 1992; Vershuren et al., 2002). In Panama's Lake Gatun, introduced peacock bass *(Cichla ocellaris)* depleted much of the native fish fauna, including small topminnows that potentially controlled mosquito larvae, thereby increasing human risk to malaria (Zaret and Paine, 1973).

Terrestrial Ecosystems

Top-down forcing and trophic cascades have been more difficult to observe and demonstrate in terrestrial ecosystems because large predators have been extirpated from so much of the world's terrestrial realm and because terrestrial plants often have such long generation times. Even so, the accumulated body of evidence is diverse and convincing.

The least equivocal evidence for terrestrial trophic cascades comes from the experimental manipulation of small predators. Removals of spiders from old-field systems in the northeastern United States (Schmitz, 2006, 2008), birds from deciduous North American trees (Marquis and Whelan, 1994), and lizards from small Bahamian islets (Schoener and Spiller, 1996, 1999) demonstrate clearly the indirect effects of predators on plants in the terrestrial realm. Numerous studies have found strong limiting effects of irrupting or high-density populations of large herbivores or seed predators on plant assemblages (Schmitz et al., 2000; Schmitz, 2008; Terborgh and Estes, 2010). Such effects, which are often caused by the loss of large predators, are known from subarctic to tropical biomes in the Old World, the New World, Australia, and many of the larger oceanic and landbridge islands (Terborgh and Estes, 2010; Estes et al., 2011).

Studies by John Terborgh and colleagues on recently created landbridge islands of the Lago Guri impoundment in Venezuela established the essential role of predation and top-down forcing in tropical forest systems. Those islands that were too small to support resident predators, and too isolated to be reached by predators residing on the mainland or on larger islands, experienced population irruptions of a diverse herbivore guild, including howler monkeys, leafcutter ants, and various rodents. Increasing herbivory resulted in high rates of seedling and sapling mortality, *recruitment* failure in many plant species, and a landscape-level shift from dense forests to parklands with a reduced overstory and practically no subcanopy (Terborgh et al., 2006).

Similar processes occur widely in temperate and boreal forests in which cervids (deer, elk, moose, and caribou) are the principal large herbivores. Following the extirpation of wolves, grizzly bears, and cougars from the United States, cervids and beavers have altered patterns of forest regeneration, markedly reducing the diversity of herbaceous plant communities and exposing stream banks to increased erosion (McShea et al., 1997; Ripple and Beschta, 2006; Waller and Rooney, 2008). Similar impacts of overabundant ungulates have been documented in other parts of the developed world that now lack large carnivores (Ripple et al., 2010). These effects are best known from various U.S. national parks, where the loss of large predators a few decades ago has left a characteristic signal of reduced growth rate of trees (McLaren and Peterson, 1994) or recruitment failure (Beschta and Ripple, 2009; Kauffman et al., 2010; Ripple et al., 2010) in the dominant tree species.

Clear ecological signals from the loss of predators and changes in top-down forcing processes have emerged over the past several decades from Sub-Saharan Africa. Studies by Justin Brashares and colleagues document a rise in baboon numbers with the loss of large predatory mammals (lion, leopard, wild dog, and cheetah) from Ghanan national parks over the past five decades. The baboons, apparently released from limitation by predation, have not only increased in abundance but have taken to eating birds and small mammals, thereby also driving those species' populations downward (Brashares et al., 2010). In equatorial East Africa, the introduction of the infectious viral disease rinderpest from Ethiopia in the late 1800s created a pandemic in wild and domestic ungulates, driving their populations downward and causing a shift in the savanna landscape from grasslands to scrub forests and bushlands (Sinclair et al., 2010). A disease control program, instituted in the 1950s, eradicated rinderpest from East Africa, thus allowing the great herds of wildebeest, buffalo, and other ungulates to recover, in turn transforming the landscape from bushlands back to grasslands and savannas. Although large predators are a conspicuous element of the East African fauna, their ecological role in this system is poorly understood. That has begun to change with work by Adam Ford and colleagues (Ford et al., 2014), who examined the interplay among large carnivores (leopard and wild dog), impalas, and acacia trees in an East African savanna. These authors discovered that the predators caused impalas to aggregate and spend most of their time in low-risk habitats, within which a thorny, and thus well-defended, acacia species *(Acacia etbaica)* predominated. The high-risk habitats, by contrast, were avoided by impalas and dominated by the less thorny, and thus more poorly defended, *A. brevispica.*

A substantial body of comparative and experimental research has established the importance of top-down forcing and trophic cascades in various terrestrial arctic and subarctic ecosystems. Lauri Oksanen and colleagues have shown that the exclusion of predatory birds and mammals from Fennoscandian Low Arctic scrublands causes small herbivorous mammals (voles and lemmings) to increase, thus leading to reductions in plant biomass and the loss of erect, woody vegetation (Aunapuu et al., 2008). In their natural configuration, these Low Arctic systems operate as classic, three-level trophic cascades. As predicted by the Fretwellian theory of food-chain length (Fretwell, 1987), the intensity of herbivory in these Low Arctic systems is chronically low. Predators are far less common in High Arctic tundra systems, owing in large measure to their intrinsically lower rates of primary production. These High Arctic systems thus operate as two-level trophic cascades. Herbivorous mammals are relatively abundant, and the intensity of herbivory is chronically high.

Productive Low Arctic scrublands appear to be dynamically comparable to forests: predatory mammals and birds regulate herbivores, and their exclusion leads to severe reduction of plant biomass and to the elimination of erect woody plants, regardless of their palatability. The new dominants are herbaceous and trailing woody plants (Aunapuu et al., 2008; Dahlgren et al., 2009). Similar plants prevail on the tundra proper. This view is supported by quantitative empirical studies and experiments (Batzli et al., 1980; Oksanen, 1983; Moen et al., 1993; Aunapuu et al., 2008; Olofsson et al., 2009). A likely reason for the inability of predators to regulate the herbivores of the tundra proper is its low primary productivity, creating a situation where the vegetation cannot sustain herbivore densities high enough to support predators (Oksanen et al., 1981; Oksanen and Oksanen, 2000).

Sergey Zimov and colleagues have argued that the present-day absence of strong grazing pressure in Beringia (northeastern Siberia and interior Alaska) is a consequence of the overkill of big arctic mammals at the end of the Pleistocene, which changed a previously graminoid-dominated "arctic steppe" to the moss–lichen–dwarf shrub tundra that occurs in the region today (Zimov et al., 1995). This view is supported by recent changes in the composition of the vegetation in areas where wild reindeer–caribou and muskoxen have recovered. Zimov and colleagues predict that if these two species of large herbivores were allowed to recover more extensively, the entire Beringian biome would transition from a moss- and lichen-dominated tundra to a grass-dominated steppe. The ecological importance of megaherbivores (such as the extinct wooly mammoth) and the system's often abundant large predators (wolves and grizzly bears) remains unknown or highly uncertain.

———————

FIFTEEN

Retrospection

I BEGAN WRITING THIS BOOK with three broad purposes. One was to recount what I had learned from nearly 50 years of studying the ecological interplay between sea otters and coastal marine ecosystems. The second was to use what I had seen and learned over the years to tell a larger story of how predators and prey interact with one another to define much of nature's operational infrastructure. My third purpose was to explain how the science really happened—how questions arose and the ways in which I strove to answer those questions. In this chapter, I'll highlight the conceptual high points of my journey.

HIGH POINT 1: THE IMPORTANCE
OF PERTURBATIONS

Each spring I teach an undergraduate class in general ecology at UC Santa Cruz. In the first lecture, I broadly define ecology as the study of the distribution and abundance of species. I then tell the students that a description of these two parameters is relatively simple, at least in principle—all one needs to do is look at nature and measure what one sees. I ask them to look outside and imagine how they might measure these parameters in the redwood forests that adorn the surrounding landscapes of our campus. The students recognize that while redwood trees are the most conspicuous feature of these landscapes, redwood forests contain numerous other species—and part of their challenge would be learning to identify these species. But that in itself doesn't seem to strike most of them as especially daunting.

I then ask them to think about why the distribution and abundance of these species are as they are, and how they might go about answering that question. I ask them to imagine different possible scenarios for just the redwood trees—that if temperature and moisture profiles differed enough, the redwoods might not live there; that if a fire swept across the hillside 50 years ago, it may have burned the less fire-tolerant plant species, thereby enhancing redwoods; that if loggers had not cut down the large, old-growth trees, the forest might be quite different from what they see today; and that if grizzly bears and mountain lions had not been exterminated or depleted from the surrounding mountains, the many deer that roam our campus might be less common, in turn influencing seedling survival—either directly from reduced grazing pressure or indirectly through the interactions of redwood seedlings with other plant species.

I go on to tell my students that these and many other interactions among species and their physical environments are what determine the distribution and abundance of species. I draw a circle on the board with dots around the perimeter to represent species, and I characterize the species' interactions by drawing lines across the circle connecting the various dots. I then ask them to think about how they might go about understanding how this infrastructure of interactions among species and their environment influences the distribution and abundance of species in the redwood forest. Some of the students immediately understand where I'm heading with all this: knowing that an interaction pathway exists tells us little or nothing, in itself, about the dynamic consequences of the interaction. I explain to them how many ecologists have attempted to solve this problem: understanding the dynamic consequences of species interactions by perturbing the distribution and abundance of species. From this point, I need only give them an example of how that was done (I use Justin Brashares's work on the "baboonification" of Africa—see chapter 14, box 2) to solidify my argument for the importance of perturbations in the study of ecology, a point I return to again and again throughout the class.

It's easy for me to argue with conviction for the importance of perturbations in ecological research because perturbations underlie both my own attempts to understand nature and so much of the work done by others I admire. But there is also a substantial challenge to perturbing nature: How does one go about doing it? The approach is straightforward in many instances. One need only add or remove a species from a study plot and measure how the distribution and abundance of other species change in response.

Results of such purposeful experimental manipulations can be embraced with high levels of confidence through replication and control. This approach has taught us a great deal about the workings of nature. But that understanding comes mostly from studies of smaller, less mobile, more abundant, and shorter-lived species. The experimental approach doesn't work very well for large apex predators, whose very natures are antithetical to the constraints of purposeful experimental manipulations.

On the surface, such constraints would seem to leave one with two choices in any quest for understanding the ecological roles of large predators: embracing broad generalizations from species and systems that can be studied experimentally, or somehow trying to figure it out from studies of unperturbed systems. The first choice requires a leap of faith that many ecologists have been unwilling to make; the second choice is intrinsically blind to dynamic process; and thus neither choice offers a way of understanding the ecological roles of the large apex predators that many ecologists find very compelling. My way around these difficulties has been to use human-caused perturbations as "natural" experiments to probe some of nature's process that cannot be seen in undisturbed, static systems and could never be recreated through purposeful manipulations.

The knock against such experiments is that they are unreplicated and uncontrolled. But with a little creative thought and a fair amount of effort, my colleagues and I have been able to replicate the removal and addition of sea otters to reef ecosystems at interisland to interregional scales across the North Pacific rim. Our "controls" are time-series measurements from places in which sea otter abundance remained more or less constant. Together with these controls, the replicated contrasts of systems with and without sea otters established the species' keystone ecological role beyond reasonable doubt. This simple approach of using large-scale disturbances as perturbations to unveil the ecological roles of large apex predators is the first conceptual high point in my evolving view of nature.

HIGH POINT 2: GENERALITY AND VARIATION

While contrasts between systems with and without apex predators provide interesting insight into the ecological roles of these animals, published reports have left some ecologists with a sense of discomfort. The discomfort seems to arise not so much from the proposed ecological processes as from the implied scales of inference. Scientists are suspicious by nature, and their suspicion is that published findings of the effects of apex predators involved some "cherry picking" of the places that the authors of these accounts have chosen to look for predator effects. The skeptics do have a point. Nature is vast and variable, and generalizing the results of a study done in some particular place to that vastness and variation will always be questionable.

I confronted this knotty issue of generality by randomly sampling the seafloor over large areas in places where the status of sea otter populations were known. My colleagues and I have repeated this sampling approach at numerous places and times. The sea otter effect does not lead to precise point estimates of species abundance, but it does lead to predictable ecosystem phase states: the uniform scarcity of sea urchins and high but variable kelp abundance where otters are present, and the uniform scarcity of kelp and high but variable sea urchin abundance where otters are absent. Surely there are shallow reefs at places across the Pacific Rim that do not fit this descriptive mold. But virtually all the many reefs I have studied do fit the mold, and thus I am confident in claiming that the sea otter–sea urchin–kelp trophic cascade is a broadly generalizable process from at least southern British Columbia to Russia's Commander Islands. The second conceptual high point in my view of nature is a generality of process. I see this generality not only within the sea otter–kelp forest system, but in a look across our planet's many other large predators and their associated ecosystems.

HIGH POINT 3: STATE SHIFTS AND HYSTERESIS

The distinctly different states that characterize ecosystems with and without sea otters do not change gradually from one to the other as otter densities wax or wane, but instead change in abrupt shifts. My colleagues and I have been able to document these state shifts on only a few occasions—in those rare places where otter populations were expanding or declining, and where we also had long enough time series of seafloor survey data to capture the event. Yet I can infer the generality of this state-shift dynamic from the absence of intermediate states in our extensive descriptions of reef systems that span the gamut of possible sea otter densities. This pattern of two distinct phase states occurs not only across the North Pacific but in kelp forest systems throughout the world (Steneck et al., 2002). In the Aleutian Islands, where I have watched otter populations both grow and collapse over the decades, I have seen differing trajectories in the functional relationship between reef phase state and sea otter density with the differing directionalities of change (i.e., hysteresis). Experimental studies done at boundaries between the kelp and urchin phase states have shown us that the fundamental nature of algal–urchin interactions changes from agonistic (beneficial to urchins and detrimental to kelp) to amensalistic (detrimental to urchins and neutral to kelp) as the system transitions from an urchin barren to a kelp forest, thus helping explain the phase shifts and hysteresis. Similar dynamics are known or suspected in other systems. So the third conceptual high point in my view of nature is the occurrence of phase shifts and hysteresis.

HIGH POINT 4: FAR-REACHING INFLUENCES OF TROPHIC CASCADES

Observing and understanding phase shifts and hysteresis in the otter–urchin–kelp trophic cascade was conceptually and methodologically challenging. Imagining and then testing for indirect effects of the otter–urchin–kelp trophic cascade on other species and ecological processes was easier. I began work in this area by reminding myself what kelp forest ecologists had long known: that kelp forests, with their high biomass and extreme productivity, are key controlling elements of coastal ecosystems. Radical differences in kelp abundance between systems with and without sea otters thus led my colleagues and me to imagine effects of the otter–urchin-cascade extending widely through coastal ecosystems. We looked for these effects just as I had looked for the effects of sea otters on sea urchins and kelp, by assembling an *interaction web* topology and then testing for the predicted effects of sea otters within that topology by contrasting otherwise similar systems with and without sea otters. In doing this, my colleagues and I have discovered numerous indirect effects of the otter–urchin–kelp trophic cascade: production effects on growth rates of filter-feeding mussels and barnacles; dietary effects on co-occurring glaucous-winged gulls and bald eagles; population effects on kelp forest finfish and sea stars; and even geochemical effects on atmospheric carbon and ocean pH. The fourth conceptual high point in my view of nature is thus the widespread and far-reaching influences of predators on the structure and function of their associated ecosystems.

Most modern biologists see the living world as a product of some 3 billion years or more of evolution by natural and sexual selection. Although indirect species interactions have been important in shaping the details, most of the immense body of work in evolutionary ecology has focused on the consequences of natural selection through direct interactions. From early on in my study of kelp forest communities, I saw strong influences of herbivory and competition on the distribution and abundance of various kelp species. I recognized that some species succeeded by being superior competitors, others by being superior defenders against their herbivores, and that the trade-off between these two strategies was the cost of defense. Peter Steinberg and I discovered what others had discovered or imagined before us: that this cost was incurred through the biosynthesis of toxic or deterrent chemical compounds. Based largely on the assumed cost of chemical defenses and the discovery of phlorotannins as toxic or deterrent in kelps and other brown algal groups, we imagined plants and their herbivores being shaped by an evolutionary "arms race" in which plants were constantly under selection to better defend themselves and herbivores were constantly under selection to better resist those defenses. I saw the apparent manifestations of these processes in the kelp forests at Amchitka Island, where a poorly defended but competitively superior flora occurred in the largely urchin-free shallow-water habitats while a well-defended but competitively inferior flora occurred in the more urchin-abundant deep-water habitats. I reasoned that sea otters, through the physiological cost of diving, created this depth-related environmental gradient in the intensity of herbivory, so it was an easy next step to imagine how the plants and their herbivores might have evolved in the absence of sea otters and their recent ancestors. I knew that the sea otter's evolutionary radiation had been confined to the North Pacific basin and, thus, I looked to Southern Hemisphere kelp forests as a place to see how the coevolution of plants and their herbivores played out in an otter-free environment. I went to New Zealand with the expectation of finding marine plants that were better defended and herbivores that were better able to resist those defenses than their respective North Pacific counterparts. And that is precisely what I found. I concluded that the top-down effects of predation by sea otters and their recent ancestors broke the selective landscape for an evolutionary arms race between marine plants and their herbivores in the

North Pacific Ocean, thus explaining the North Pacific's poorly defended marine flora, the tendency of North Pacific kelp forests to collapse by way of distinct phase shifts following the loss of sea otters, and a host of reciprocal interactions with species like abalones and sea cows that fed on kelp. These various discoveries and observations have led to the fifth conceptual high point in my view of nature: the evolutionary consequences of indirect species interactions.

HIGH POINT 6: INTER-ECOSYSTEM CONNECTIVITY

At the beginning of my career as an ecologist and naturalist, I thought of ecosystems largely as self-contained functional entities. That perspective was based more on convenience than on anything else. I have since come to realize that coastal ecosystems are connected with the land and the open sea by various processes, including those associated with the movements of large animals. My first glimpse of this was through the episodic spawning migration of smooth lumpsuckers from somewhere in the oceanic realm to coastal waters of the Aleutian Islands, an event that provided winter food subsidy to the sea otters and broke the typical late winter–early spring pattern of population regulation through starvation. My second glimpse of cross-ecosystem linkages was through the interplay among introduced foxes, native seabirds, and the seabirds' vectoring of nutrients from a rich ocean to an impoverished land. The effect of foxes on island ecosystems, which drove a change in the latter from highly productive maritime grasslands to less productive maritime tundra, occurred by way of the sea. My final and most poignant glimpse of inter-ecosystem connectivity was through the search for a cause of the collapse of sea otters and coastal ecosystems in southwest Alaska, which began in the early 1990s and continues to this day. I discovered killer whale predation as the probable cause of the sea otter's population collapse and thought that an event in the open sea must have driven otherwise more oceanic killer whales to the coastal zone to begin eating sea otters. Further research led my colleagues and me to conclude that the killer whale–sea otter interaction was only a small part of the story. The interactive network, as we see it, began with post–World War II industrial whaling in the North Pacific Ocean, swept through seals and sea lions during the ensuing decades, and eventually found its way to sea otters and kelp forests in the adjacent coastal oceans some 50 years later. Thus, the sixth conceptual high point to my view of nature is the functional linking together of structurally distinct ecosystems through the movements of large animals.

HIGH POINT 7: SERENDIPITY

As a young person looking forward to a career in ecology, I began to develop a long view of how I imagined my life unfolding. A few of those early visions were realized, but most were not. This is because my career has been guided by serendipitous events that I could never have foreseen. I almost certainly would not have become a marine biologist had I not failed my draft physical in 1970. A chance meeting with Bob Paine in 1971 opened my mind to a view of ecological process that defined the rest of my career. Had that chance meeting not happened, I probably would have taken a different path. My chance association with Peter Steinberg when he was a graduate student at UC Santa Cruz led to our work on plant–herbivore coevolution, which has strongly influenced my view of the workings of kelp forest ecosystems ever since. Had Peter chosen to study elsewhere, my view of kelp forest ecosystems would surely be very different today. Our veterinarian's fiancé's tragic farming accident caused us to tag sea otters in Clam Lagoon on Adak Island, a default decision that later allowed me to reject a number of seemingly reasonable explanations for the sea otter's decline and conclude that the likely cause was killer whale predation. Had that accident not happened, we would not have tagged sea otters in Clam Lagoon, in which case the evidence would not have been strong enough to allow us to come forward with the killer whale hypothesis. My chance reunion with Vernon Byrd in the U.S. Fish and Wildlife Service's bunkhouse at Adak led to the fox–seabird project. Had the ship departed earlier or had Vernon and I missed one another for any number of possible reasons, I would never have pursued the fox–seabird work. I could go on for pages with other examples. The point is that my career pathway was not defined by a long look forward, but by chance events, most of which surely would be missed if I could rerun my life's clock. Therefore, the seventh conceptual high point from my life as a naturalist is the importance of serendipity. Young ecologists near the beginnings of their careers may find this unsettling. I'll bet that those of you who succeed will look back on your careers with the same view I now hold of how and why my own happened.

My life as a naturalist has been driven by a love of nature and the excitement of learning about how nature works. My modus operandi as a scientist has been to recognize serendipitous events for the opportunities they provided, and to make the best of those opportunities. I've always felt confident in my ability recognize these events and to learn from them. But I also admit to occasional feelings of self-doubt and more frequent uncertainty over how both my approach to science and the things I thought I had learned would be viewed by others. These doubts were assuaged not only by my own sense of truth, but by various salutations and awards from my colleagues in the scientific community. In 2011, I received the Western Society of Naturalists Lifetime Achievement Award. In 2012, I received the American Society of Mammalogists C. Hart Merriam Award for lifetime excellence in research. These prestigious awards, coming in back-to-back years, helped me understand that the many flaws and soft points in my life's work were overshadowed by the impacts of my ideas and discoveries. I had always thought of myself as a competent scientist with the creative ability to ask interesting questions and the personal wherewithal to follow up with the field studies and analyses needed to answer those questions. But I never thought of myself as elite or truly gifted. I was therefore overwhelmed by emotion and pride when, at 6:05 AM on April 27, 2014, I was awakened by a phone call informing me that I had been elected to the National Academy of Sciences. That honor helped me realize that, in the minds of some of the world's finest and most respected scientists, my accomplishments are seen as substantial and perhaps even groundbreaking.

Acknowledgments and awards reflect other people's perspectives on my work, not my own. I'm proud of what I've accomplished, but I still can't help but feel as though my accomplishments were somehow realized by my being swept along by the hand of fate. As time passes, it's easy to turn uncertainties into truths. I think I got much of it right along the way, though I may have gotten some of it wrong and I'm certain that I missed important details. My life as a naturalist has been exciting and joyful, at least for the most part. I can't imagine anything more challenging, more humbling, and more important to the future welfare of our planet than the quest to understand nature. In looking back now, just a year short of 70, I have indeed been blessed with a wonderful life.

———

Looking to the Future

EVERYTHING I'VE WRITTEN TO THIS POINT reflects the past in one way or another. This final short chapter is a look to the future. I'll begin with my own thoughts on where the science of top-down forcing and trophic cascades should go from here, and I'll end by discussing the implications of what we already know from that science for natural resource policy and management. I'll address these issues with the example of sea otters in particular and large predators more generally.

RESEARCH

So long as human societies value science and train scientists, scientists will argue the need for more scientific research. Yet society is frequently reluctant to act on what science already knows. Scientists and society should ask a dualistic question in considering any area of scientific research: What do we know, and what do we need to know? Future scientific understanding will not always build on our resulting templates—what we thought we needed to know and thus sought to learn. Scientists sometimes stumble onto important discoveries. But these discoveries usually begin with a quest for understanding something blessed by society as being important to know.

What is it, then, that science should strive to understand about sea otters and kelp forests? I don't believe it necessary to further document the existence of the sea otter–kelp forest trophic cascade in other places and times. We have observed and documented this interaction enough to know that it's real. That said, I believe there would be value in the more extensive documentation of the interaction's generality and variation. This has been done across the heart of the sea otter's natural range—from the western Aleutian Islands through British Columbia. It has not been done in the sea otter's eastern and western range peripheries—from Washington State to the central Pacific coast of Baja California, and from Kamchatka through the Kurile Islands to northern Japan. I'm almost certain that the relationship of the rocky-reef phase state to the presence or absence of sea otters is less well defined in California and Mexico than in Alaska and Canada. But until someone does a proper survey, the degree to which that may be true will remain unknown.

I'm of two minds on the need for further research on indirect effects of the otter-urchin-kelp trophic cascade. Finding more examples will incrementally expand the details of something we already know—that the indirect effects of sea otters extend widely through coastal ecosystems. But at some point, adding one or a few more examples becomes uninteresting. Moreover, there is almost no end to the possibilities. On the other hand, some of these as-yet-unknown indirect effects of sea otter predation may be important to science and policy. Recent discovery of a link between the otter–urchin–kelp trophic cascade and atmospheric carbon sequestration (Wilmers et al., 2012) was new and important enough to justify the effort. The fate of kelp-derived organic carbon and the link between kelp forests and other species that are valued by human societies (e.g., herring, salmon, and nearshore-feeding whales) are issues I consider worthy of further effort.

Although strong indirect effects of sea otter predation almost certainly led to evolution through natural selection, these evolutionary effects have barely been considered, much less looked for. With the recent discovery that ecological and evolutionary changes proceed rapidly in lockstep (Schoener, 2011; Thompson, 2013), further study of *eco-evo reciprocity* in the sea otter–kelp forest system is an exciting new frontier. There is evidence that sea otters and their recent ancestors modified the coevolution of defense and resistance in North Pacific kelps and their herbivores, and that the resulting effect on kelp nutritional quality has further shaped the evolution of various kelp consumers (see chapter 9). Other strong direct and indirect interactions extend broadly from sea otters through coastal ecosystems (see chapter 8). Some of these interactions almost surely became similarly entangled in eco-evo reciprocity. Such evolutionary dimensions to nature are admittedly difficult to observe and demonstrate. But they must be there, parts of living nature that remain hidden, waiting for some clever scientist to look for and find them.

Top-down forcing and trophic cascades have now been documented in virtually every major ecosystem type on the planet (Terborgh and Estes, 2010). Our early account of sea otters and kelp forests (Estes and Palmisano, 1974), earlier lake studies by Brooks and Dodson (Brooks and Dodson, 1965) and others, and the earliest published account of wolf–ungulate–tree interactions (McLaren and Peterson, 1994) were novel, surprising, and perhaps even eye-popping to many in the ecological community. How many more case studies are needed to establish the general importance of these processes in nature? None, in my opinion. There is value in reporting novel trophic cascades, but that value is no longer in the discovery of something that is new and exciting. I must be in a minority in this view, because novel examples continue to be published in journals like *Science* and *Nature*. Be that as it may, the more interesting quest should be for systems in which large-predator effects do not occur. The discovery of such a system would cause me to take note with real interest and wonder. While I'm sincere in saying this, I also recognize the inherent inferential difficulties in demonstrating that something does not exist, and thus I would never urge a young scientist to look specifically for such a thing.

What, then, do we need to know about other predators and their ecosystem-level effects? Two important goals are a better understanding of mechanism and better documentation of generality and variation. Predators can affect their prey through what have been referred to (obtusely, I believe) as *density-mediated* (consumptive) and *trait-mediated* (fear-induced, nonconsumptive) effects. Although trait-mediated effects clearly occur in nature, the relative importance of these two potential mechanisms of predator-induced effects on their prey has become a topic of considerable interest and debate. The wolf–elk–aspen system in Yellowstone is an interesting case in point. Bill Ripple and Bob Beschta have argued that fear of being eaten by wolves has caused elk to avoid risky habitats, thus promoting aspen recovery in such places (Ripple and Beschta, 2004). Critics have challenged this interpretation by pointing out that elk kill locations do not correlate very well with so-called risky habitats (Kauffman et al., 2010). Both arguments have important weaknesses. Little direct evidence has been gathered of trait-mediated effects of wolves on elk—which is not to say that such effects don't occur, only that they haven't been clearly demonstrated. But carcass locations don't resolve the issue very well, because wolves are "coursing" predators, which means that where an elk is killed is not indicative of where the kill began. This particular debate is probably resolvable through more detailed behavioral–ecological studies, as, for example, have recently been done on the interplay among large predatory mammals, impalas, and acacia trees in East Africa (Ford et al., 2014). Impalas avoid risky places, and that avoidance behavior influences plant–herbivore interactions. I'll bet that when all is said and done, we will see that many species interact in the same general ways.

Although top-down forcing and trophic cascades have been widely demonstrated across global ecosystems, the question of generality and variation in these interactions is largely unresolved. There are various reasons why. One is that scientists haven't addressed the question in a careful and rigorous way. A more rigorous assessment of generality and variation will require two essential elements in future research protocols. The first is to define a geographic area (the statistical sample space or statistical population) to which the inference will be made. This will be tricky, especially for terrestrial systems with their high degree of inherent spatial heterogeneity. The sample space might be defined by species ranges (the predator, for example) or habitat types (aspen forests, for example). Then, after the sample space has been defined, the evidence must be gathered from a set of representative locations within that sample space. Until such analyses are done in some reasonable manner, critical scientists will quibble (and properly so) about generality and variation. From my corner, this is the central difficulty of the debate over wolves and trophic cascades in Yellowstone. Recently published analyses by Luke Painter and colleagues (Painter et al., 2014) that chronicle changing aspen recruitment dynamics over large areas of northern Yellowstone strike me as an important step in the right direction.

A related challenge to future understanding of top-down forcing and trophic cascades is the incorporation of appropriate scales of time in the way these dynamics should be expected to play out. Trophic cascades have been easy to see in lakes because they occur quickly, owing largely to the short generation times of the autotrophs, herbivores, and predators. By the same logic, trophic cascades will be difficult to see and document for systems in which the central players have long generation times. Trophic cascades will be inherently difficult to see and demonstrate in real time for any system in which long-lived trees are the autotrophs. Other approaches, such as the historical tree-ring analyses used by McLaren and Peterson (McLaren and Peterson, 1994) on Isle Royale, and by Ripple and colleagues in various U.S. national parks (Ripple and Larsen, 2000; Ripple and Beschta, 2007), will be needed for a proper understanding of trophic cascades and top-down forcing.

Many other interesting dimensions to top-down forcing and trophic cascades undoubtedly remain to be explored and discovered. But those identified in the preceding paragraphs will be needed to provide a base of knowledge for most of them.

MANAGEMENT AND POLICY

I chose a career with the federal government in part because I thought it would provide a direct conduit from my research to natural resource conservation and management. In retrospect, this has been one of my greatest disappointments. Early on, the U.S. Fish and Wildlife Service (USFWS) used the sea otter–kelp forest trophic cascade as a small part of their argument for why California sea otters needed to expand in numbers and range. But in fact, most of their policy and management decisions were based on demographic issues, not ecological ones. I don't blame them for this, because their mandates were largely defined and dictated by the language of the Endangered Species Act, which is vague and scientifically outdated on the ecological issues (Soulé et al., 2005).

The USFWS's current behavior in Alaska is more troubling. The sea otters that were reintroduced to southeast Alaska in the late 1960s and early '70s have increased spectacularly, to a present overall population of about 25,000 animals. And there is room for further population growth. As these animals spread into previously unoccupied habitats, they are reducing sea urchins and enhancing kelp forests. The otters are also depleting various shellfish stocks— which, in turn, is affecting local fisheries. State and federal politicians are pressuring the USFWS to manage this problem. Exactly how the service should do that isn't clear. They will probably never get away with supporting local culls, because by doing that they would almost certainly be sued for violation of the Endangered Species and Marine Mammal Protection acts. The USFWS has yet to establish a policy or figure out what to do. But their public dialogue has tended strongly toward somehow helping local communities keep their shellfishing grounds free of sea otters. The USFWS, in my view, needs to establish a policy that considers the science they have supported over the years, not just the cost of lost shellfisheries. Although that cost should be part of their decision-making process, it should also be weighed against the benefits of enhanced kelp forests.

Similar issues surround the conservation and management of wolves in North America. As wolves have recovered and spread into various midwestern and western states, the USFWS has turned wolf management over to the states, which by and large want to limit and control wolves to protect domestic and wild ungulates. Neither the USFWS nor the states seem to be weighing the wolf–ungulate–tree trophic cascade into their management policies. It appears to me as though the USFWS, an agency that once held broad views on the needs of natural resource conservation and management, has regressed to their century-old mandate of predator control.

Looking beyond this immediate situation, what does our current understanding of top-down forcing and trophic cascades imply about the future of natural resource management and conservation? Everyone needs to recognize that a world without large predators will be a very different place than one in which these animals are retained. The detrimental effects to people from fish, wildlife, and livestock losses are the most obvious and well-known part of that equation. But the ecological roles of large predators extend far beyond these direct losses.

No sensible person would advocate the complete repatriation of large predators everywhere. A global policy is needed that considers the broader ecological importance of these animals and weighs the costs and benefits of their effects on nature and society. Consider, for example, the growing human population and land-based food production. A future world will almost certainly have to rely more on the direct human consumption of plants if it is to sustain current and forecast human populations. In moving toward that end, the costs of living with predators may be far outweighed by their ecological benefits.

Living with large predators means rethinking the spatial scales of conservation and management. Small protected areas simply won't do, as they cannot maintain viable populations of large predators, especially large predatory mammals. We will need landscapes and seascapes that are managed and protected at large enough scales to maintain these animals at ecologically effective population densities.

A few forward-looking conservation groups, such as the Wildlands Network, understand this need and are striving to reconnect nature at the necessary scales. The Spine of the Continent project, whose goal is to link up various public and private lands in the Intermountain West from Mexico to the Yukon, is one such highly ambitious example. Whether or not these efforts will succeed remains uncertain. There are daunting obstacles, perhaps the most challenging of which is public education on the roles and values of predators.

To my thinking, the Wildlands vision is clear and true. Any serious effort to conserve biodiversity must be accompanied by the conservation of large carnivores. But first things first. Instead of bickering over the details, will scientists embrace this view? And if so, will the public ever accept it?

ACKNOWLEDGMENTS

Although I had considered writing a book for some years, my decision to finally do so crystallized during a seminar visit to Cornell University in September 2005. Harry Greene and his then graduate student Josh Donlan hosted me, and I flew out to Ithaca early to spend a day in the field with them. Inspired by the beauty and majesty of two large eastern diamondback rattlesnakes (admittedly located through their implanted radio transmitters), our discussions turned to Harry's highly successful book on snakes. Knowing something of the content and thread of my research on sea otters and kelp forests, he urged me to give it a go. And Josh encouraged me not to wait too long. I was less than a month short of my sixtieth birthday, and Josh's advice resonated in me with a sense of urgency. Although seven more years would pass before I settled on a concept and began writing, I thank Harry and Josh for that initial nudge.

Looking further back in time, I was inspired, encouraged, and helped by various key people. My uncle, Frank Springer, instilled in me a love for science and nature, a sense of curiosity, and a passion for romance and adventure at a young age. Bill Reiner's undergraduate ecology class at the University of Minnesota focused my interest in biology on nature and natural history. Norm Smith provided friendship, support, and guidance during my years as a graduate student at the University of Arizona. Clyde Jones taught me to believe in myself and encouraged me to strive for scientific excellence. Ancel Johnson, my first supervisor as a federal research scientist, gave me the freedom to explore. Bob Paine turned my budding view of ecology in a productive direction by urging me to think about the importance of predation. Paul Dayton helped me, during intellectually insecure early years, with friendship, humor, critical evaluations of my ideas and work, and thoughtful advice. Mary Power encouraged me with her enthusiasm for the things I was learning and inspired me with her own excellent ecological research. I remember her telling me during a particularly dark period, "Those people [referring to skeptics and critics] may not get it, but the undergraduate students in my ecology course at Berkeley do." Michael Soulé opened my mind to new perspectives on science, self, and why people behave as they do. John Terborgh, the best naturalist I've ever known, helped me understand that the things I learned about sea otters and kelp forests occur widely in nature. I will forever be grateful to all these people.

While this book is a personal account of my life's journey in science, I had help and company along the way. For contributions to the science, I am indebted to Bob Anthony, Corinne Bacon, Doug Burn, Don Croll, Eric Danner, Dan Doak, David Duggins, Matt Edwards, Tom Gelatt, Jim Gilbert, Krista Hanni, David Irons, Ron Jameson, Walter Jarman, David Jessup, Brenda Konar, Kristin Laidre, David Lindberg, John Maron, Melissa Miller, Mike Murray, John Palmisano, Galen Rathbun, Stacey Reese, Shauna Reisewitz, Marianne Riedman, Charles Simenstad, Don Siniff, Alan Springer, Michelle Staedler, Peter Steinberg, Bob Steneck, Jim Taggart, Gus van Vliet, Ken Vicknair, Jane Watson, Jon Watt, Terrie Williams, Chris Wilmers, and Cindy Zabel. I am particularly grateful to Tim Tinker for his friendship and our close intellectual partnership over the past two decades.

For assistance in the field, I thank Jack Ames, Brenda Bellachy, Gena Bentall, Jim Bodkin, Stacey Buckalew, Heather Colletti, Bob Cowen, Jim Coyer, Angie Doroff, George Esslinger, Tom Evans, Jared Figurski, Mike Foster, Chris Harrold, Brian Hatfield, Sandy Hawes, David Irons, Ladd Johnson, Mike Kenner, Kim Kloeker, Carolyn Kurle, Bob Mayer, Carolyn McCormick, Keith Miles, Kathy Ann Miller, Dan Monson, Mike Murray, Dan Reed, Elaine Rhode, Mark Ricca, Alex Rose, Mike Russell, Diana Steller, Julie Stewart, Glenn VanBlaricom, Ben Weitzman, and Frank Winter. Brian Hatfield and Mike Kenner deserve particular thanks for dedicated help with diving and hands-on work with sea otters during the past three decades. I apologize to at least a few others who inevitably slipped through the cracks of my memory of work in the field over the past 50 years. To all, I am grateful not only for your able help, but for sharing the joy of time spent together in wild and wondrous places.

Fieldwork in remote areas requires financial and logistical support. Funding for my work during the earlier years was provided by the Research Division of the U.S. Fish and Wildlife Service. From about the mid-1980s onward, my research was supported by contracts and grants from the U.S. National Science Foundation, the U.S. Navy, the U.S. Air Force, and the North Pacific Research Board. Field support for work in the Aleutian Islands was provided by the Alaska Maritime National Wildlife Refuge, especially through efforts by Vernon Byrd, Dan Boone, Lisa Spitler, and Jeff Williams. I am grateful to the U.S. Coast Guard for transporting equipment and field personnel from Kodiak to Adak, Amchitka, Attu, and Shemya islands during routine supply and surveillance missions. Fieldwork elsewhere in the Aleutians required ship support, provided over the years by the M/V *Aleutian Tern*, R/V *Alpha Helix*, M/V *Tiglax*, R/V *Thomas G. Thompson*, M/V *Norseman*, and R/V *Point Sur*. I am especially grateful to captains George Putney, Kevin Bell,

Billy Pepper, and Paul Tate for safe transport through coastal waters of southwest Alaska, occasionally during inclement conditions.

The Department of Ecology and Evolutionary Biology and the Institute of Marine Sciences at the University of California, Santa Cruz, have been a physical and intellectual home throughout most of my career. I thank Bill Doyle, Gary Griggs, Maria Choy, Kay House, Susan Thuringer, and Kathy Durcan for administrative support. I am grateful to faculty and students, especially Mark Carr, Dan Costa, Don Croll, Dan Doak, Laurel Fox, Burney LeBoeuf, Bruce Lyon, John Pearse, Pete Raimondi, and Chris Wilmers, for inspiration and intellectual companionship. Terrie Williams, my wife, colleague, and life companion for the past quarter-century, deserves special thanks for her humor, support, and ideas—and for tolerating my long periods away from home.

The manuscript was read in its entirety or in part by Christina Eisenberg, Harry Greene, Bob Paine, Mary Power, Erin Rechsteiner, Anne Salomon, Alan Springer, Peter Steinberg, Terrie Williams, and an anonymous reviewer. Although I am grateful for their comments, suggestions, and corrections, I admit to not heeding all of their advice, and I alone am responsible for what now appears in print.

No life is lived in a single dimension. I thank my wonderful children, Colin and Anna, for their love, support, and the incomparable richness they have added to my journey.

GLOSSARY

ACTIVITY BUDGET Proportion of time an animal spends in principal life-support activities such as foraging, resting, and grooming.

ALEUTS Aboriginal peoples of the Aleutian Islands and Alaska Peninsula.

ALGIVORE A consumer that feeds on algae.

ALTERNATIVE STABLE STATE One of two or more ecosystem configurations that can occur in nature under identical physical and biological conditions.

AMENSALISM A species interaction in which one species incurs a cost and the other species is unaffected (–/o).

APEX PREDATOR [or apex consumer] The top species in a food chain.

APPARENT COMPETITION An indirect interaction between a predator and two prey in which an increase in one of the prey species enhances the predator's numbers, thus leading to a decline in the second prey species.

ASSIMILATION EFFICIENCY The proportion of a consumer's diet that it digests and assimilates.

AUTOTROPH An organism capable of self-nourishment by using inorganic materials as a source of nutrients and using photosynthesis or chemosysthesis as a source of energy.

BETA DIVERSITY That element of overall landscape-level species diversity (or gamma diversity) attributable to differences between habitats (e.g., hilltops vs. valleys, north slopes vs. south slopes).

BIOGEOGRAPHY The study of the distribution and abundance of species over large scales of space and time.

BIOTA The assemblage of all organisms.

BIOTOXIN A poisonous substance produced by a living organism (as opposed to toxins produced by humans).

BOTTOM-UP FORCING [or bottom-up control] The process by which the distribution and abundance of species are dictated by production and by the efficiency of energy and material flux upward through food webs.

BYCATCH Inadvertent capture of nontarget species in a fishery.

CARBON CYCLE The biogeochemical cycle by which carbon is exchanged among the biosphere, geosphere, pedosphere (soils), hydrosphere, and atmosphere.

CARRYING CAPACITY The maximum population size that can be supported by the environment.

CETACEANS The group (Order) of mammals that includes whales, dolphins, and porpoises.

COEVOLUTION The process by which two interacting organisms evolve together, each changing as a result of changes in the other.

COMMENSALISM A species interaction in which one species benefits and the other is unaffected (+/o).

COMPETITION A species interaction in which each species incurs a cost (−/−).

CONSUMER A species that feeds on other species.

CONTEXT DEPENDENCY The sometimes qualitatively different way in which two species interact with one another, depending on ecological context.

CONTINENTAL ISLANDS Islands that occur on continental shelves.

CORIOLIS EFFECT The deflection of moving objects over Earth's surface, to the right in the Northern Hemisphere and to the left in the Southern Hemisphere.

DEMOGRAPHY The study of population-controlling processes.

DENSITY-DEPENDENCE The processes whereby reproductive and mortality rates vary with population density.

DENSITY-MEDIATED Adjective denoting ecological influences of predators caused by the killing of their prey.

DETRITIVORE A consumer that feeds on the breakdown products of already dead organisms.

DEW LINE Distant Early Warning Line, a system of radar stations designed to detect incoming Soviet bombers to North America during the Cold War.

DIRECT EFFECT An influence of one species on another in which there are no other, interceding species.

ECO-EVO RECIPROCITY The interplay between ecology and evolution.

ECOLOGICAL DRIVER An ecological process that causes a change in the distribution and/or abundance of a species.

ECOLOGICAL PASSENGER A species whose distribution and/or abundance is caused to change by an ecological driver.

EKMAN TRANSPORT Deflected transport of ocean surface waters due to wind and the Coriolis effect. This surface water movement is typically from onshore to offshore in the eastern North Pacific Ocean.

EL NIÑO–SOUTHERN OSCILLATION Anomalously warm or cold surface seawater temperatures that develop in the tropical western Pacific and spread eastward and poleward along the west coast of North America.

ENDANGERED SPECIES ACT A U.S. federal law designed to prevent extinction of critically imperiled species.

EUPHOTIC ZONE Water depths in lakes and oceans below which net primary production becomes zero and then negative.

EUROPEAN CARBON MARKET An exchange market for carbon led by the European Union.

EUTROPHIC Nutrient-rich (e.g., "a eutrophic environment").

FATHOM A measure of water depth (1 fathom = 6 feet).

FIELD METABOLIC RATE The metabolic rate (joules per gram body mass per time) of a free-living animal in the wild.

FOCAL ANIMAL SAMPLING Obtaining population characteristics by the integration of longitudinal information from individual animals.

FOOD CHAIN A group of species, ranging from autotrophs to apex consumers, that are linked together through consumer–prey interactions.

FOOD WEB A network of interconnected food chains.

FORAGING THEORY A body of theory that provides a guide to decision rules by foraging animals, based largely on the associated benefits of energy gain and the costs of the forager itself being eaten.

FUNCTIONAL GROUP A collection of organisms based on some unifying character (e.g., trophic level).

GADIDS A family of fish (Gadidae) that includes cod and their relatives.

GLACIAL AGE A period of Earth history during which temperatures were abnormally cool.

HAUL OUT Behavioral tendency for animals like pinnipeds and sea otters to move from water to ice or land.

HERBIVORE An organism that consumes plants.

HOMEOTHERM [adjective: homeothermic] An organism, such as a bird or mammal, whose body temperature is constant and largely independent of its environmental temperature.

HOME RANGE The area over which an animal lives and travels during the course of its life.

HYSTERESIS A functional relationship between ecological drivers and passengers that varies depending on the directionality of change in the abundance or intensity of the driver.

INDIRECT EFFECT An influence of one species on another in which there is one or more interceding species.

INTERACTION STRENGTH The quantitative influence of one species on the abundance of another.

INTERACTION WEB The network of interactions among species and between species and their physical environments.

INTERNATIONAL WHALING COMMISSION (IWC) An international body created by the terms of the International Convention for the Regulation of Whaling (ICRW), intended to provide for the proper conservation and management of whale stocks.

INTERTIDAL ZONE Shoreline area between the water marks of spring high and low tides.

INVASIBILITY The degree to which an ecosystem or habitat is capable of being invaded by an exotic species.

ISOTOPE Forms of chemical elements that differ in the number of neutrons. Isotopic analyses are commonly used to study the source of materials in nature.

KELP Marine macroalgae in the order Laminariales.

KEYSTONE SPECIES Any species that is comparatively rare in nature but imparts a strong effect on its associated ecosystem.

LAMBDA (λ) Finite rate of population increase; the proportional change in population size from one year to the next.

LORAN A long-range navigation system that was used widely throughout the world from the late 1950s to the early twenty-first century.

LOTKA-VOLTERRA EQUATIONS A series of equations that describe the joint population dynamics of predators (consumers) and their prey (victims).

LYME DISEASE An infectious disease caused by *Borrelia* (a bacterium) and contracted through tick bites.

MACROALGAE Kelp and other larger multicellular algae that commonly live attached to the seafloor.

MATRILINE An intergenerational pedigree of mothers to their female offspring.

MIDDEN A mound or deposit containing shells, animal bones, and other refuse.

MINERALIZATION The process by which organic substances are transformed into inorganic substances.

MIOCENE EPOCH Period of Earth history from about 23 million to 5 million years ago.

MUTUALISM A species interaction in which both species benefit (+/+).

NAUTICAL MILE A unit of distance approximately equal to one minute of latitude (= 1.852 kilometers).

NEAP TIDES Moderate high and low tides that occur around the first and third quarters of the moon.

NEOTROPICAL Pertaining to the New World tropics.

NET PRIMARY PRODUCTION Gross primary production minus respiration. Measured as mass of carbon per area per time.

OCEANIC ISLANDS Islands that occur beyond the continental shelves, most of which are created by deep-sea volcanism.

OLIGOTROPHIC Adjective denoting environments that are nutrient-impoverished.

OPTIMAL FORAGING THEORY A theory of dietary choice based on both the abundance of a consumer's potential prey species and the value of those prey to the consumer. Optimal foraging theory generally predicts increased dietary diversity with reduced abundance of the most valuable (preferred) prey.

PACIFIC DECADAL OSCILLATION (PDO) A multidecadal pattern of temperature change in the Pacific Ocean. Warm and cold phases of the PDO are characterized by punctuated transitions.

PCB Acronym for polychlorinated biphenyl, a synthetic organic compound built around two benzene rings.

PERCENT COVER The percentage of habitat (land or seafloor) covered by some particular organism or group of organisms.

PERTURBATION A disturbance of any kind to an otherwise static system.

PERTURBATION EXPERIMENT Purposeful perturbations used to test hypothesized effects of an ecological driver.

PHASE SHIFT An abrupt change in ecosystem structure in response to a graded change in the ecological driver.

PHASE STATE One of two or sometimes more distinctly different structural configurations of a biological community or ecosystem.

PHLOROTANNINS Organic molecules composed of 1,3,5 tri-hydroxy benzene units (phloroglucinol) linked together by ester or carbon–carbon bonds.

PHYLOGENETIC TREE The evolutionary branching pattern of any group of related organisms, beginning with the basal species.

PHYLOGENY The evolutionary history of relatedness for a particular group of organisms.

PHYLOGEOGRAPHY Study of historical processes responsible for current distribution of organisms.

PHYTOPLANKTON Unicellular algae that fuel water-column primary production in aquatic ecosystems.

PINNIPEDS The group (Order) of mammals that includes seals, sea lions, and walruses.

PISCIVOROUS Feeding on fish.

PLANKTIVOROUS Feeding on plankton in the water column.

PLEISTOCENE EPOCH Period of Earth history from about 2.5 million to 12,000 years ago.

PLIOCENE EPOCH Period of Earth history from about 5 million to 2.5 million years ago.

POPULATION STRUCTURE Frequency of occurrence of differing age (age structure) or size (size structure) classes within a population.

PREY A species that is fed upon by some other species.

PRIMARY PRODUCTION Rate of carbon fixation by autotrophs through photosynthesis or chemosynthesis.

PRIORITY EFFECT The influence of the prior occurrence of a species on subsequent community development or change.

QUADRAT A square frame used to sample plants and animals.

RECRUITMENT The addition of recently born, juvenile, or young organisms to a population.

SAMPLE SPACE The population being sampled.

SCAN SAMPLES Population characteristics obtained from the integration of instantaneous information from multiple individuals.

SCAT Fecal material, commonly from a mammal carnivore.

SEA STARS Multiarmed predatory invertebrates (Echinodermata; Asteroidea) that live on the seafloor.

SEA URCHINS Spine-covered herbivorous invertebrates (Echinodermata; Echinoidea) that live on the rocky seafloor and are eaten by sea otters.

SECONDARY METABOLITES Organic compounds produced by an organism but not directly involved in that organism's normal development, growth, or reproduction.

SECONDARY PRODUCTION Rate of carbon fixation by heterotrophs through consumption and assimilation of prey.

SPACE FOR TIME An approach to the study of long-term phenomena (e.g., succession) wherein observations from different places at which the phenomenon began at differing times are used to reconstruct the phenomenon's time course.

SPRING TIDES Extreme high and low tides that occur around full and new moons.

STATE SPACE A graph in which the abundances of two species (or groups of organisms) are plotted on the x and y axes.

SUBTIDAL ZONE Area below water mark of spring low tides.

TEST Calcified exoskeleton of a sea urchin.

TIME SERIES A series of repeated measurements, typically of the same thing, obtained through time.

TOP-DOWN FORCING [or top-down control] The process by which the distribution and abundance of species are dictated by the consequences of consumer-induced mortality downward through food webs.

TRAIT-MEDIATED Adjective denoting influences of predators caused by the fear of being eaten.

TRANSARCTIC INTERCHANGE Movement of marine organisms across the Arctic between the Pacific and Atlantic Ocean basins.

TROPHIC CASCADE Indirect interactions between consumers and their prey that extend downward (from higher to lower trophic levels) through food chains or food webs.

TROPHIC LEVEL The position of a species in a food chain or food web, in relation to that of the autotrophs.

ZOOPLANKTON Small animals that live in the water columns of marine and freshwater ecosystems.

BIBLIOGRAPHY

Alexander, R. (1974). The evolution of social behaviour. *Annual Review of Ecology, Evolution, and Systematics, 5,* 325–383.

Ames, J. A., Hardy, R. A., & Wendell, F. E. (1983). Tagging materials and methods for sea otters, *Enhydra lutris. California Fish and Game, 69,* 243–252.

Anderson, C. B., Pastur, G. M., Lencinas, M. V., Wallem, P., Moorman, M. C., & Rosemond, A. D. (2009). Do introduced North American beavers engineer differently in southern South America? An overview with implications for restoration. *Mammal Review, 39,* 33–52.

Anderson, C. B., Rozzi, R., Torres-Mura, J. C., McGehee, S. M., Sherriffs, M. F., Schuettler, E., & Rosemond, A. D. (2006). Exotic vertebrate fauna in the remote and pristine sub-Antarctic Cape Horn Archipelago region of Chile. *Biodiversity and Conservation, 15,* 3295–3313.

Anthony, R. G., Estes, J. A., Ricca, M. A., Miles, A. K., & Forsman, E. D. (2008). Bald eagles and sea otters in the Aleutian archipelago: Indirect effects of trophic cascades. *Ecology, 89,* 2725–2735.

Aunapuu, M., Dahlgren, J., Oksanen, T., Grellmann, D., Oksanen, L., Olofsson, J., et al. (2008). Spatial patterns and dynamic responses of arctic food webs corroborate the Exploitation Ecosystems Hypothesis (EEH). *American Naturalist, 171,* 249–262.

Babcock, R. C., Kelly, S., Shears, N. T., Walker, J. W., & Willis, T. J. (1999). Changes in community structure in temperate marine reserves. *Marine Ecology Progress Series, 189,* 125–134.

Bacon, C. E., Jarman, W. M., Estes, J. A., Simon, M., & Norstrom, R. J. (1999). Comparison of organochlorine contaminants among sea otter *(Enhydra lutris)* populations in California and Alaska. *Environmental Toxicology and Chemistry, 18,* 452–458.

Bailey, E. (1995). Introduction of foxes to Alaskan islands—history, effects on avifauna, and eradication. U.S. Fish and Wildlife Service Technical Report 193 (p. 117).

Barkai, A., & McQuaid, C. (1988). Predator–prey role reversal in a marine benthic ecosystem. *Science, 242,* 62–64.

Batzli, G. O., White, R. G., McLean, S. F., Pitelka, F. A., & Collier, B. D. (1980). The herbivore-based trophic system. In J. Brown, P. C. Miller, & F. Bunnell (Eds.), *An Arctic Ecosystem: The Coastal Tundra at Barrow, Alaska* (pp. 335–410). Strouds-burg, PA: Dowden, Hutchinson & Ross.

Baum, J. K., & Worm, B. (2009). Cascading top-down effects of changing oceanic predator abundances. *Journal of Animal Ecology, 70,* 699–714.

Ben-David, M., Bowyer, R. T., Duffy, L. K., Roby, D. D., & Schell, D. M. (2008). Social behavior and ecosystem processes: River otter latrines and nutrient dynam-ics of terrestrial vegetation. *Ecology, 79,* 2567–2571.

Berger, J., Stacey, P. B., Bellis, L., & Johnson, M. P. (2001). A mammalian preda-tor–prey imbalance: Grizzly bear and wolf extinction affect avian Neotropical migrants. *Ecological Applications, 11,* 947–960.

Berger, K. M., Gese, E. M., & Berger, J. (2008). Indirect effects and traditional trophic cascades: A test involving wolves, coyotes, and pronghorn. *Ecology, 89,* 818–828.

Beschta, R. L., & Ripple, W. J. (2006). River channel dynamics following extirpa-tion of wolves in northwestern Yellowstone National Park, USA. *Earth Surface Processes and Landforms, 31,* 1525–1539.

Beschta, R. L., & Ripple, W. J. (2009). Large predators and trophic cascades in ter-restrial ecosystems of the western United States. *Biological Conservation, 142,* 2401–2414.

Brashares, J., Prugh, L. R., Stoner, C. J., & Epps, C. W. (2010). Ecological and con-servation implications of mesopredator release. In J. Terborgh & J. A. Estes (Eds.), *Trophic Cascades: Predators, Prey and the Changing Dynamics of Nature* (pp. 221–240). Washington, DC: Island Press.

Breen, P. A., Carson, T. A., Foster, J. B., & Stewart, E. A. (1982). Changes in subtidal community structure associated with British Columbia sea otter transplants. *Marine Ecology Progress Series, 7,* 13–20.

Brooke, M. de L., & Davies, N. B. (1988). Mimicry by cuckoos *Cuculus canorus* in relation to discrimination by hosts. *Nature, 335,* 630–631.

Brooks, J. L., & Dodson, S. I. (1965). Predation, body size, and composition of plank-ton. *Science, 150,* 28–35.

Burkepile, D. E., & Hay, M. E. (2006). Herbivore vs. nutrient control of marine primary producers: Context-dependent effects. *Ecology, 87,* 3128–3139.

Burton, A. (2010). Let's pray for the lamprey. *Frontiers in Ecology and the Environ-ment, 8,* 392.

Carpenter, S. R., Cole, J. J., Hodgson, J. R., Kitchell, J. F., Pace, M. L., Bade, D., et al. (2001). Trophic cascades, nutrients, and lake productivity: Whole-lake experi-ments. *Ecological Monographs, 71,* 163–186.

Carpenter, S. R., Kitchell, J. F., & Hodgson, J. R. (1985). Cascading trophic interac-tions and lake productivity. *BioScience, 35,* 634–649.

Carpenter, S. R., Kitchell, J. F., Hodgson, J. R., Cochran, P., Elser, J. J., Elser, M. M., et al. (1987). Regulation of lake primary productivity by food web structure. *Ecol-ogy, 68,* 1863–1876.

Carter, J., Foote, A. L., & Johnson-Randall, L. A. (1999). Modeling the effects of nutria *(Myocastor coypus)* on wetland loss. *Wetlands, 19*, 209–219.

Casini, M., Hjelm, J., Molinero, J.-C., Lövgren, J., Cardinale, M., Bartolino, V., et al. (2009). Trophic cascades promote threshold-like shifts in pelagic marine ecosystems. *Proceedings of the National Academy of Sciences USA, 106*, 197–202.

Casini, M., Lövgren, J., Hjelm, J., Cardinale, M., Molinero, J. C., & Kornilovs, G. (2008). Multi-level trophic cascades in a heavily exploited open marine ecosystem. *Proceedings of the Royal Society B, 275*, 1793–1801.

Cowen, R. K. (1983). The effects of sheephead *(Semicossyphus pulcher)* predation on red sea urchin *(Strongylocentrotus franciscanus)* populations: An experimental analysis. *Oecologia, 58*, 249–255.

Croll, D. A., Kadula, R., & Tershy, B. R. (2006). Ecosystem impacts of the decline of large whales in the North Pacific. In J. A. Estes, D. P. DeMaster, D. F. Doak, T. M. Williams, & R. Brownell (Eds.), *Whales, Whaling and Ocean Ecosystems* (pp. 202–214). Berkeley: University of California Press.

Croll, D. A., Maron, J. L., Estes, J. A., Danner, E. M., & Byrd, G. V. (2005). Introduced predators transform subarctic islands from grassland to tundra. *Science, 307*, 1959–1961.

Crooks, K., & Soulé, M. (1999). Mesopredator release and avifaunal extinctions in a fragmented system. *Nature, 400*, 563–566.

Dahlgren, J., Oksanen, L., Olofsson, J., & Oksanen, T. (2009). Plant defenses at no cost? The recovery of tundra scrubland following heavy grazing by gray-sided voles *(Myodes rufocanus)*. *Evolutionary Ecology Research, 11*, 1205–1216.

Daskalov, G. M., Grishin, A. N., Rodinov, S., & Mihneva, V. (2007). Trophic cascades triggered by overfishing reveal possible mechanisms of ecosystem regime shifts. *Proceedings of the National Academy of Sciences USA, 104*, 10518–10523.

Davenport, A. C., & Anderson, T. W. (2007). Positive indirect effects of reef fishes on kelp performance: The importance of mesograzers. *Ecology, 88*, 1548–1561.

Dayton, P. K. (1975). Experimental studies of algal canopy interactions in a sea otter–dominated kelp community at Amchitka Island, Alaska. *Fishery Bulletin, 73*, 230–237.

DeMaster, D. P., Trites, A. W., Clapham, P., Mizroch, S., Wade, P., Small, R. J., & Ver Hoef, J. (2006). The sequential megafaunal collapse hypothesis: Testing with existing data. *Progress in Oceanography, 68*, 329–342.

Doak, D. F., Estes, J. A., Halpern, B. S., Jacob, U., Lindberg, D. R., Lovvorn, J., et al. (2008). Understanding and predicting ecological dynamics: Are major surprises inevitable? *Ecology, 89*, 952–961.

Domning, D. P. (1978). Sirenian evolution in the North Pacific Ocean. *University of California Publications in Geological Sciences, 118*, 176.

Doroff, A. M., Estes, J. A., Tinker, M. T., Burn, D. M., & Evans, T. J. (2003). Sea otter population declines in the Aleutian archipelago. *Journal of Mammalogy, 84*, 55–64.

Duggins, D. (1980). Kelp beds and sea otters: An experimental approach. *Ecology, 61*, 447–453.

Duggins, D., Simenstad, C. A., & Estes, J. A. (1989). Magnification of secondary production by kelp detritus in coastal marine ecosystems. *Science, 245*, 170–173.

Dulvy, N. K., Freckleton, R. P., & Polunin, N. V. C. (2004). Coral reef cascades and the indirect effects of predator removal by exploitation. *Ecology Letters, 7*, 410–416.

Dworjanyn, S. A., Wright, J. T., Paul, N. A., De Nys, R., & Steinberg, P. D. (2006). Cost of chemical defence in the red alga *Delisea pulchra. Oikos, 113*, 13–22.

Emslie, S. D., & Patterson, W. P. (2007). Abrupt recent shift in $\delta^{13}C$ and $\delta^{15}N$ values in Adélie penguin eggshell in Antarctica. *Proceedings of the National Academy of Sciences USA, 104*, 11666–11669.

Essington, T. (2010). Trophic cascades in open ocean ecosystems. In J. Terborgh & J. A. Estes (Eds.), *Trophic Cascades: Predators, Prey and the Changing Dynamics of Nature* (pp. 91–105). Washington, DC: Island Press.

Estes, J. A. (1977). Population estimates and feeding behavior of sea otters. In M. L. Merritt and R. G. Fuller (Eds.), *The Environment of Amchitka Island* (pp. 511–526). TID-26712. Springfield, VA: USERDA.

Estes, J. A. (1990). Growth and equilibrium in sea otter populations. *Journal of Animal Ecology, 59*, 385–401.

Estes, J. A., Brashares, J. S., & Power, M. E. (2013). Predicting and detecting reciprocity between indirect ecological interactions and evolution. *American Naturalist, 181*, S76–S99.

Estes, J. A., DeMaster, D. P., Doak, D. F., Williams, T. M., & Brownell, R. L. J. (Eds.). (2006). *Whales, Whaling and Ocean Ecosystems.* Berkeley: University of California Press.

Estes, J. A., Doak, D. F., Springer, A. M., & Williams, T. M. (2009a). Causes and consequences of marine mammal population declines in southwest Alaska: A food-web perspective. *Philosophical Transactions of the Royal Society B, 364*, 1647–1658.

Estes, J. A., Doak, D. F., Springer, A. M., Williams, T. M., & van Vliet, G. B. (2009b). Trend data do support the sequential nature of pinniped and sea otter declines in the North Pacific Ocean, but does it really matter? *Marine Mammal Science, 25*, 748–754.

Estes, J. A., & Duggins, D. O. (1995). Sea otters and kelp forests in Alaska: Generality and variation in a community ecological paradigm. *Ecological Monographs, 65*, 75–100.

Estes, J. A., & Gilbert, J. R. (1978). Evaluation of an aerial survey of Pacific walruses *(Odobenus rosmarus divergens). Journal of the Fisheries Research Board of Canada, 35*, 1130–1140.

Estes, J. A., Jameson, R. J., & Rhode, E. B. (1982). Activity and prey selection in the sea otter: Influence of population status on community structure. *American Naturalist, 120*, 242–258.

Estes, J. A., Lindberg, D. R., & Wray, C. (2005). Evolution of large body size in abalones *(Haliotis):* Patterns and implications. *Paleobiology, 31*, 591–606.

Estes, J. A., & Palmisano, J. F. (1974). Sea otters: Their role in structuring nearshore communities. *Science, 185,* 1058–1060.

Estes, J. A., Peterson, C. H., & Steneck, R. S. (2010a). Some effects of apex predators in higher-latitude coastal oceans. In J. Terborgh & J. A. Estes (Eds.), *Trophic Cascades: Predators, Prey and the Changing Dynamics of Nature* (pp. 37–54). Washington, DC: Island Press.

Estes, J. A., & Steinberg, P. D. (1988). Predation, herbivory, and kelp evolution. *Paleobiology, 14,* 19–36.

Estes, J. A., Terborgh, J., Brashares, J. S., Power, M. E., Berger, J., Bond, W. J., et al. (2011). Trophic downgrading of planet Earth. *Science, 333,* 301–306.

Estes, J. A., Tinker, M. T., & Bodkin, J. L. (2010b). Using ecological function to develop recovery criteria for depleted species: Sea otters and kelp forests in the Aleutian archipelago. *Conservation Biology, 24,* 852–860.

Estes, J. A., Tinker, M., Williams, T. M., & Doak, D. F. (1998). Killer whale predation on sea otters linking oceanic and nearshore ecosystems. *Science, 282,* 473–476.

Flecker, A. S., & Townsend, C. R. (1994). Community-wide consequences of trout introduction in New Zealand streams. *Ecological Applications, 4,* 798–807.

Ford, A. T., Goheen, J. R., Otieno, T. O., Bidner, L., Isbell, L. A., Palmer, T. M., et al. (2014). Large carnivores make savanna tree communities less thorny. *Science, 346,* 346–349.

Foster, M. S., & Schiel, D. R. 1988. Kelp communities and sea otters: Keystone species or just another brick in the wall? In G. R. VanBlaricom and J. A. Estes (Eds.), *The Community Ecology of Sea Otters* (pp. 92–115). Ecological Studies 65. New York: Springer.

Frank, K. T., Petrie, B., Choi, J. S., & Leggett, W. C. (2005). Trophic cascades in a formerly cod-dominated ecosystem. *Science, 308,* 1621–1623.

Frank, K. T., Petrie, B., Fisher, J. A. D., & Leggett, W. C. (2011). Transient dynamics of an altered large marine ecosystem. *Nature, 477,* 86–89.

Fretwell, S. D. (1987). Food chain dynamics: The central theory of ecology? *Oikos, 50,* 291–301.

Friedlander, A. M., & Parrish, J. D. (1998). Temporal dynamics of fish communities on an exposed shoreline in Hawaii. *Environmental Biology of Fishes, 53,* 1–18.

Fuller, D. A., Sasser, C. E., Johnson, W. B., & Gosselink, J. G. (1985). The effects of herbivory on vegetation on islands in Atchafalaya Bay, Lousiana. *Wetlands, 4,* 105–114.

Garfield, B. (2010). *The Thousand Mile War: World War II in Alaska and the Aleutian Islands* (4th ed.). Fairbanks: University of Alaska Press.

Gelatt, T. S., Siniff, D. B., & Estes, J. A. (2002). Activity patterns and time budgets of the declining sea otter population at Amchitka Island, Alaska. *Journal of Wildlife Management, 66,* 29–39.

Gruner, D. S. (2005). Biotic resistance to an invasive spider conferred by generalist insectivorous birds on Hawai'i Island. *Biological Invasions, 7,* 541–546.

Hairston, N. G., Smith, F. E., & Slobodkin, L. B. (1960). Community structure, population control, and competition. *American Naturalist, 94,* 421–425.

Hanni, K. D., Mazet, J. A. K., Gulland, F. M. D., Estes, J., Staedler, M., Murray, M. J., et al. (2003). Clinical pathology and assessment of pathogen exposure in southern and Alaskan sea otters. *Journal of Wildlife Diseases, 39,* 837–850.

Heck, K. L., Jr., Pennock, J. R., Valentine, J. F., Coen, L. D., & Sklenar, S. A. (2000). Effects of nutrient enrichment and small predator density on seagrass ecosystems: An experimental assessment. *Limnology and Oceanography, 45,* 1041–1057.

Hixon, M. (1991). Predation as a process structuring coral reef fish communities. In P. F. Sale (Ed.), *The Ecology of Fishes on Coral Reefs.* San Diego, CA: Academic Press.

Hixon, M. A., & Beets, J. P. (1993). Predation, prey refuges, and the structure of coral-reef fish assemblages. *Ecological Monographs, 63,* 77–101.

Hockey, P. A., & Branch, G. (1984). Oystercatchers and limpets: Impact and implications. A preliminary assessment. *Ardea, 72,* 199–206.

Holdo, R. M., Sinclair, A. R. E., Dobson, A. P., Metzger, K. L., Bolker, B. M., Ritchie, M. E., & Holt, R. D. (2009). A disease-mediated trophic cascade in the Serengeti and its implications for Ecosystem C. *PLoS Biology, 7*(9), e1000210.

Holt, R. D. (1977). Predation, apparent competition, and the structure of prey communities. *Theoretical Population Biology, 12,* 197–229.

Hrbaček, J., Dvorakova, M., Korinek, V., & Prochazkova, L. (1961). Demonstration of the effects of the fish stock on the species composition of zooplankton and the intensity of metabolism of the whole plankton assemblage. *Verhandlungen: Internationale Vereinigung für Theoretischeund Angewand Te Limnologie, 14,* 192–195.

Hughes, T. P. (1994). Catastrophes, phase shifts, and large-scale degradation of a Caribbean coral reef. *Science, 265,* 1547–1551.

Hughes, T. P., Rodrigues, M. J., Bellwood, D. R., Ceccarelli, D., Hoegh-Guldberg, O., McCook, L., et al. (2007). Phase shifts, herbivory, and the resilience of coral reefs to climate change. *Current Biology, 17,* 360–365.

Irons, D. B., Anthony, R. G., & Estes, J. A. (1986). Foraging strategies of glaucous-winged gulls in a rocky intertidal community. *Ecology, 67,* 1460–1474.

Jackson, J. B., Kirby, M. X., Berger, W. H., Bjorndal, K. A., Botsford, L. W., Bourque, B. J., et al. (2001). Historical overfishing and the recent collapse of coastal ecosystems. *Science, 293,* 629–637.

Jameson, R. J., Kenyon, K. W., Johnson, A. M., & Wight, H. M. (1982). History and status of translocated sea otter populations in North America. *Wildlife Society Bulletin, 10,* 100–107.

Jessup, D. A., Johnson, C. K., Estes, J., Carlson-Bremer, D., Jarman, W. M., Reese, S., et al. (2010). Persistent organic pollutants in the blood of free-ranging sea otters (*Enhydra lutris* ssp.) in Alaska and California. *Journal of Wildlife Diseases, 46,* 1214–1233.

Jicha, B. R., Scholl, D. W., Singer, B. S., Yogodzinski, G. M., & Kay, S. M. (2006). Revised age of Aleutian Island Arc formation implies high rate of magma production. *Geology, 34,* 661–664.

Jones, R. D. (1965). Sea otters in the Near Islands, Alaska. *Journal of Mammalogy,* *46,* 702.

Kauffman, M. J., Brodie, J. F., & Jules, E. S. (2010). Are wolves saving Yellowstone's aspen? A landscape-level test of a behaviorally mediated trophic cascade. *Ecology,* *91,* 2742–2755.

Kaufman, L. (1992). Catastrophic change in species-rich freshwater ecosystems. *BioScience, 42,* 846–858.

Kenyon, K. W. (1969). Sea otters in the eastern North Pacific Ocean. *North American Fauna, 68,* 1–352.

Kirby, M. X. (2004). Fishing down the coast: Historical expansion and collapse of oyster fisheries along continental margins. *Proceedings of the National Academy of Sciences USA, 107,* 12163–12167.

Knight, T. M., McCoy, M. W., Chase, J. M., McCoy, K. A., & Holt, R. D. (2005). Trophic cascades across ecosystems. *Nature, 437,* 880–883.

Kuker, K., & Barrett-Lennard, L. (2010). A re-evaluation of the role of killer whales *Orcinus orca* in a population decline of sea otters *Enhydra lutris* in the Aleutian Islands and a review of alternative hypotheses. *Mammal Review, 40,* 103–124.

Kvitek, R. G., Iampietro, P. J., & Bowlby, C. E. (1998). Sea otters and benthic prey communities: A direct test of the sea otter as keystone predator in Washington State. *Marine Mammal Science, 14,* 895–902.

Lafferty, K. D. (2004). Fishing for lobsters indirectly increases epidemics in sea urchins. *Ecological Applications, 14,* 1566–1573.

Levi, T., Kilpatrick, A. M., Mangel, M., & Wilmers, C. C. (2012). Deer, predators, and the emergence of Lyme disease. *Proceedings of the National Academy of Sciences USA, 109,* 10942–10947.

Ling, S. D., Johnson, C. R., Frusher, S. D., & Ridgway, K. R. (2009). Overfishing reduces resilience of kelp beds to climate-driven catastrophic phase shift. *Proceedings of the National Academy of Sciences USA, 106,* 22341–22345.

Magurran, A. E. (1990). The adaptive significance of schooling as an antipredator defense in fish. *Annales Zoologici Fennici, 27,* 51–66.

Markel, R. M., & Shurin, J. B. (in press). Indirect effects of sea otters on rockfish (*Sebastes* spp.) in giant kelp forests. *Ecology.* doi:10.1890/14–0492.1

Marquis, R. J., & Whelan, C. J. (1994). Insectivorous birds increase growth of white oak through consumption of leaf-chewing insects. *Ecology, 75,* 2007–2014.

Marshall, K. N., Hobbs, N. T., & Cooper, D. J. (2013). Stream hydrology limits recovery of riparian ecosystems after wolf reintroduction. *Proceedings of the Royal Society B, 280,* 20122977.

McCauley, D. J., DeSalles, P. A., Young, H. S., Dunbar, R. B., Dirzo, R., Mills, M. M., & Micheli, F. (2012). From wing to wing: The persistence of long ecological interaction chains in less-disturbed ecosystems. *Scientific Reports, 2.*

McClanahan, T. R., Sala, E., Stickels, P. A., Cokos, B. A., Baker, A. C., Starger, C. J., & Jones, S. H. (2003). Interaction between nutrients and herbivory in controlling algal communities and coral condition on Glover's Reef, Belize. *Marine Ecology Progress Series, 261,* 135–147.

McFarland, W. N. (1991). The visual world of coral reef fishes. In P. F. Sale (Ed.), *The Ecology of Fishes on Coral Reefs.* San Diego, CA: Academic Press.

McIntosh, A. R., & Townsend, C. R. (1996). Interactions between fish, grazing invertebrates and algae in a New Zealand stream: A trophic cascade mediated by fish-induced changes to grazer behaviour? *Oecologia, 108,* 174–181.

McLaren, B. E., & Peterson, R. O. (1994). Wolves, moose, and tree rings on Isle Royale. *Science, 266,* 1555–1558.

McShea, W. J., Underwood, H. B., & Rappole, J. H. (1997). *The Science of Overabundance.* Washington, DC: Smithsonian Institution Press.

Miller, K. A., & Estes, J. A. (1989). Western range extension for *Nereocystis leutkeana* in the North Pacific Ocean. *Botanica Marina, 32,* 535–538.

Mizroch, S. A., & Rice, D. W. (2006). Have North Pacific killer whales switched prey species in response to depletion of the great whale populations? *Marine Ecology Progress Series, 310,* 235–246.

Moen, J., Lundberg, P. A., Ekerholm, P., & Oksanen, L. (1993). Lemming grazing on snowbed vegetation during a population peak, northern Norway. *Arctic and Alpine Research, 25,* 130–135.

Monson, D. H., Estes, J. A., Bodkin, J. L., & Siniff, D. B. (2000). Life history plasticity and population regulation in sea otters *(Enhydra lutris). Oikos, 90,* 457–468.

Moore, J. W., Schindler, D. E., Carter, J. L., Fox, J., & Holtgrieve, G. W. (2007). Biotic control of stream fluxes: Spawning salmon drive nutrient and matter export. *Ecology, 88,* 1278–1291.

Mumby, P. J., Dahlgren, C. P., Harborne, A. R., Kappel, C., Micheli, F., Brumbaugh, D., et al. (2006). Fishing, trophic cascades, and the process of grazing on coral reefs. *Science, 311,* 98–101.

Murie, O. J. (1959). Fauna of the Aleutian Islands and Alaska Peninsula. *North American Fauna, 61,* 1–364.

Myers, R. A., Baum, J. K., Shepherd, T. D., Powers, S. P., & Peterson, C. H. (2007). Cascading effects of the loss of apex predatory sharks from a coastal ocean. *Science, 315,* 1846–1850.

Naiman, R. J., Bilby, R. E., Schindler, D. E., & Helfield, J. M. (2002). Pacific salmon, nutrients, and the dynamics of freshwater and riparian ecosystems. *Ecosystems, 5,* 399–417.

Nelson, B. V., & Vance, R. R. (1979). Diel foraging patterns of the sea urchin *Centrostephanus coronatus* as a predator avoidance strategy. *Marine Biology, 51,* 251–258.

Newman, M. H., Paredes, G. A., Sala, E., & Jackson, J. B. C. (2006). Structure of Caribbean coral reef communities across a large gradient of fish biomass. *Ecology Letters, 9,* 1216–1227.

Nicol, S., Bowie, A., Jarman, S., Lannuzel, D., Meiners, K. M., & Van Der Merwe, P. (2010). Southern Ocean iron fertilization by baleen whales and Antarctic krill. *Fish and Fisheries, 11,* 203–209.

Ogden, J. C., Brown, R. A., & Salesky, N. (1973). Grazing by the echinoid *Diadema antillarum philippi:* Formation of halos around West Indian patch reefs. *Science, 182,* 715–717.

Oksanen, L. (1983). Trophic exploitation and arctic phytomass patterns. *American Naturalist, 122,* 45–52.

Oksanen, L., Fretwell, S. D., Arruda, J., & Niemelä, P. (1981). Exploitation ecosystems in gradients of primary productivity. *American Naturalist, 118,* 240–261.

Oksanen, L., & Oksanen, T. (2000). The logic and realism of the hypothesis of exploitation ecosystems. *American Naturalist, 155,* 703–723.

O'Leary, J. K., & McClanahan, T. R. (2010). Trophic cascades result in large-scale coralline algae loss through differential grazer effects. *Ecology, 91,* 3584–3597.

Olofsson, J., Oksanen, L., Callaghan, T., Hulme, P. E., Oksanen, T., & Suominen, O. (2009). Herbivores inhibit climate driven shrub expansion on the tundra. *Global Change Biology, 15,* 2681–2693.

O'Neill, D. (2007). *The Firecracker Boys: H-Bombs, Inupiat Eskimos, and the Roots of the Environmental Movement* (p. 418). New York: Basic Books.

Pace, M. L., Cole, J. J., Carpenter, S. R., & Kitchell, J. F. (1999). Trophic cascades revealed in diverse ecosystems. *Trends in Ecology and Evolution, 14,* 483–488.

Paine, R. T. (1974). Intertidal community structure: Experimental studies on the relationship between a dominant competitor and its principal predator. *Oecologia, 15,* 93–120.

Paine, R. T. (1980). Food webs: Linkage, interaction strength and community infrastructure. *Journal of Animal Ecology, 49,* 666–685.

Paine, R. T., & Vadas, R. L. (1969). The effects of grazing by sea urchins, *Strongylcentrotus* spp., on benthic algal populations. *Limnology and Oceanography, 14,* 710–719.

Painter, L. E., Beschta, R. L., Larsen, E. J., & Ripple, W. J. (2014). Recovering aspen follow changing elk dynamics in Yellowstone: Evidence of a trophic cascade? *Ecology, 96,* 252–263.

Persson, L., Diehl, S., Johansson, L., Andersson, G., & Hamrin, S. F. (1992). Trophic interactions in temperate lake ecosystems—a test of food chain theory. *American Naturalist, 140,* 59–84.

Polis, G. A., & Hurd, S. D. (1996). Linking marine and terrestrial food webs: Allochthonous input from the ocean supports high secondary productivity on small islands and coastal land communities. *American Naturalist, 147,* 396–423.

Power, M. E. (1990). Effect of fish in river food webs. *Science, 250,* 811–815.

Power, M. E., Matthews, W. J., & Stewart, A. J. (1985). Grazing minnows, piscivorous bass, and stream algae: Dynamics of a strong interaction. *Ecology, 66,* 1448–1456.

Power, M. E., Parker, M. S., & Dietrich, W. E. (2008). Seasonal reassembly of river food webs under a Mediterranean hydrologic regime: Floods, droughts, and impacts of fish. *Ecological Monographs, 78,* 263–282.

Randall, J. E. (1965). Grazing effect on sea grasses by herbivorous reef fishes in the West Indies. *Ecology, 46,* 255–260.

Rasher, D. B., & Hay, M. E. (2010). Chemically rich seaweeds poison corals when not controlled by herbivores. *Proceedings of the National Academy of Sciences USA, 107,* 9683–9688.

Reisewitz, S. E., Estes, J. A., & Simenstad, C. A. (2006). Indirect food web interactions: Sea otters and kelp forest fishes in the Aleutian archipelago. *Oecologia, 146,* 623–631.

Ripple, W. J., & Beschta, R. L. (2004). Wolves and the ecology of fear: Can predation risk structure ecosystems? *BioScience, 54,* 755.

Ripple, W. J., & Beschta, R. L. (2006). Linking a cougar decline, trophic cascade, and catastrophic regime shift in Zion National Park. *Biological Conservation, 133,* 397–408.

Ripple, W. J., & Beschta, R. L. (2007). Hardwood tree decline following large carnivore loss on the Great Plains, USA. *Frontiers in Ecology and the Environment, 5,* 241–246.

Ripple, W. J., Beschta, R. L., Fortin, J. K., & Robbins, C. T. (2014). Trophic cascades from wolves to grizzly bears in Yellowstone. *Journal of Animal Ecology, 83,* 223–233.

Ripple, W. J., & Larsen, E. J. (2000). Historic aspen recruitment, elk, and wolves in northern Yellowstone National Park, USA. *Biological Conservation, 95,* 361–370.

Ripple, W. J., Rooney, T. P., & Beschta, R. L. (2010). Large predators, deer, and trophic cascades in boreal and temperate ecosystems. In J. Terborgh & J. A. Estes (Eds.), *Trophic Cascades: Predators, Prey and the Changing Dynamics of Nature* (pp. 141–161). Washington, DC: Island Press.

Ripple, W. J., Wirsing, A. J., Beschta, R. L., & Buskirk, S. W. (2011). Can restoring wolves aid in lynx recovery? *Wildlife Society Bulletin, 35,* 514–518.

Roman, J., Estes, J. A., Morissette, L., Smith, C., Costa, D., McCarthy, J., et al. (2014). Whales as marine ecosystem engineers. *Frontiers in Ecology and the Environment, 12,* 377–385.

Roman, J., & McCarthy, J. J. (2010). The whale pump: Marine mammals enhance primary productivity in a coastal basin. *PloS ONE, 5,* e13255.

Sandin, S. A., & Pacala, S. W. (2005). Fish aggregation results in inversely density-dependent predation on continuous coral reefs. *Ecology, 86,* 1520–1530.

Sandin, S. A., Walsh, S. M., & Jackson, J. B. C. (2010). Prey release, trophic cascades, and phase shifts in tropical nearshore ecosystems. In J. Terborgh & J. A. Estes (Eds.), *Trophic Cascades: Predators, Prey and the Changing Dynamics of Nature* (pp. 71–90). Washington, DC: Island Press.

Scheffer, M., Carpenter, S., Foley, J. A., Folke, C., & Walker, B. (2001). Catastrophic shifts in ecosystems. *Nature, 413,* 591–596.

Schindler, D. E., Carpenter, S. R., Cole, J. J., Kitchell, J. F., & Pace, M. L. (2008). Influence of food web structure on carbon exchange between lakes and the atmosphere. *Science, 277,* 248–251.

Schmitz, O. J. (2006). Predators have large effects on ecosystem properties by changing plant diversity, not plant biomass. *Ecology, 87,* 1432–1437.

Schmitz, O. J. (2008). Herbivory from individuals to ecosystems. *Annual Review of Ecology, Evolution, and Systematics, 39,* 133–152.

Schmitz, O. J., Hambäck, P. A., & Beckerman, A. P. (2000). Trophic cascades in terrestrial systems: A review of the effects of carnivore removals on plants. *American Naturalist, 155,* 141–153.

Schoener, T. W. (2011). The newest synthesis: Understanding the interplay of evolutionary and ecological dynamics. *Science, 331,* 426–429.

Schoener, T. W., & Spiller, D. A. (1996). Devastation of prey diversity by experimentally introduced predators in the field. *Nature, 381,* 691–694.

Schoener, T. W., & Spiller, D. A. (1999). Indirect effects in an experimentally staged invasion by a major predator. *American Naturalist, 153,* 347–358.

Silliman, B. R., & Bertness, M. D. (2002). A trophic cascade regulates salt marsh primary production. *Proceedings of the National Academy of Sciences USA, 99,* 10500–10505.

Sinclair, A. R. E., Metzger, K., Brashares, J. S., Nkwabi, A., Sharam, G., & Fryxell, J. M. (2010). Trophic cascades in tropical savannas: Serengeti as a case study. In J. Terborgh & J. A. Estes (Eds.), *Trophic Cascades: Predators, Prey and the Changing Dynamics of Nature* (pp. 255–274). Washington, DC: Island Press.

Smith, J. E., Smith, C. M., & Hunter, C. L. (2001). An experimental analysis of the effects of herbivory and nutrient enrichment on benthic community dynamics on a Hawaiian reef. *Coral Reefs, 19,* 332–342.

Soulé, M. E., Estes, J. A., Miller, B., & Honnold, D. L. (2005). Strongly interacting species: Conservation policy, management, and ethics. *BioScience, 55,* 168–176.

Springer, A. M., Estes, J. A., van Vliet, G. B., Williams, T. M., Doak, D. F., Danner, E. M., Forney, K. M., & Pfister, B. (2003). Sequential megafaunal collapse in the North Pacific Ocean: An ongoing legacy of industrial whaling? *Proceedings of the National Academy of Sciences USA, 100,* 12223–12228.

Springer, A. M., Estes, J. A., van Vliet, G. B., Williams, T. M., Doak, D. F., Danner, E. M., & Pfister, B. (2008). Mammal-eating killer whales, industrial whaling, and the sequential megafaunal collapse in the North Pacific Ocean: A reply to critics of Springer et al. 2003. *Marine Mammal Science, 24,* 414–442.

Steinberg, P. D. (1989). Biogeographical variation in brown algal polyphenolics and other secondary metabolites: Comparison between temperate Australasia and North America. *Oecologia, 78,* 373–382.

Steinberg, P. D., Estes, J. A., & Winter, F. C. (1995). Evolutionary consequences of food-chain length in kelp forest communities. *Proceedings of the National Academy of Sciences USA, 92,* 8145–8148.

Steneck, R. S., Graham, M. H., Bourque, B. J., Corbett, D., Erlandson, J. M., Estes, J. A., & Tegner, M. J. (2002). Kelp forest ecosystems: Biodiversity, stability, resilience and future. *Environmental Conservation, 29,* 436–459.

Steneck, R. S., Vavrinec, J., & Leland, A. V. (2004). Accelerating trophic level dysfunction in kelp forest ecosystems of the western North Atlantic. *Ecosystems, 7,* 323–331.

Tegner, M. J., & Dayton, P. K. (2000). Ecosystem effects of fishing in kelp forest communities. *ICES Journal of Marine Science, 57,* 579–589.

Terborgh, J., & Estes, J. A. (Eds.) (2010). *Trophic Cascades: Predators, Prey and the Changing Dynamics of Nature.* Washington, DC: Island Press.

Terborgh, J., Feeley, K., Silman, M., Nuñez, P., & Balukjan, B. (2006). Vegetation dynamics on predator-free land-bridge islands. *Journal of Ecology, 94,* 253–263.

Thacker, R. W., Ginsburg, D. W., & Paul, V. J. (2001). Effects of herbivore exclusion and nutrient enrichment on coral reef macroalgae and cyanobacteria. *Coral Reefs, 19,* 318–329.

Thompson, J. N. (2013). *Relentless Evolution.* Chicago: University of Chicago Press.

Trites, A. W., Deecke, V. B., Gregr, E. J., Ford, J. K. B., & Olesiuk, P. F. (2007). Killer whales, whaling, and sequential megafaunal collapse in the North Pacific: A comparative analysis of the dynamics of marine mammals in Alaska and British Columbia following commercial whaling. *Marine Mammal Science, 23,* 751–765.

Vadas, R. L. (1970). Preferential feeding: An optimization strategy in sea urchins. *Ecological Monographs, 47,* 337–371.

Vershuren, D., Johnson, T. C., Kling, H. J., Edgington, D. N., Leavitt, P. R., Brown, E. T., Talbot, M. R., & Hecky, R. E. (2002). History and timing of human impact on Lake Victoria, East Africa. *Proceedings of the Royal Society B, 269,* 289–294.

Vicknair, K., & Estes, J. A. (2012). Interactions among sea otters, sea stars, and suspension-feeding invertebrates in the western Aleutian archipelago. *Marine Biology, 159,* 2641–2649.

Wade, P. R., Burkanov, V. N., Dahlheim, M. E., Friday, N. A., Fritz, L. W., Loughlin, T. R., et al. (2007). Killer whales and marine mammal trends in the North Pacific—A re-examination of evidence for sequential megafauna collapse and the prey-switching hypothesis. *Marine Mammal Science, 23,* 766–802.

Wade, P. R., Ver Hoef, J. M., & Demaster, D. P. (2009). Mammal-eating killer whales and their prey—Trend data for pinnipeds and sea otters in the North Pacific Ocean do not support the sequential megafaunal collapse hypothesis. *Marine Mammal Science, 25,* 737–747.

Waller, D. M., & Rooney, T. P. (2008). *The Vanishing Present: Wisconsin's Changing Lands, Waters, and Wildlife.* Chicago: University of Chicago Press.

Watson, J., & Estes, J. A. (2011). Stability, resilience, and phase shifts in rocky subtidal communities along the west coast of Vancouver Island, Canada. *Ecological Monographs, 81,* 215–239.

Watt, J., Siniff, D. B., & Estes, J. A. (2000). Inter-decadal patterns of population and dietary change in sea otters at Amchitka Island, Alaska. *Oecologia, 124,* 289–298.

Whitehead, H., & Reeves, R. (2005). Killer whales and whaling: The scavenging hypothesis. *Biology Letters, 1,* 415–418.

Williams, I. D., & Polunin, N. V. C. (2001). Large-scale associations between macroalgal cover and grazer biomass on mid-depth reefs in the Caribbean. *Coral Reefs, 19,* 358–366.

Williams, T. D., Williams, A. L., & Siniff, D. B. (1981). Fentanyl and azaperone produced neuroleptanalgesia in the sea otter. *Journal of Wildlife Disease, 17,* 401–404.

Williams, T. M., Estes, J. A., Doak, D. F., & Springer, A. M. (2004). Killer appetites: Assessing the role of predators in ecological communities. *Ecology, 85,* 3373–3384.

Wilmers, C. C., Estes, J. A., Edwards, M., Laidre, K. L., & Konar, B. (2012). Do trophic cascades affect the storage and flux of atmospheric carbon? An analysis of sea otters and kelp forests. *Frontiers in Ecology and the Environment, 10,* 409–415.

Wilmers, C. C., & Getz, W. M. (2005). Gray wolves as climate change buffers in Yellowstone. *PLoS Biology, 3,* 571–576.

Winter, F. C., & Estes, J. A. (1992). Experimental evidence for the effects of polyphenolic compounds from *Dictyoneurum californicum* Ruprecht (Phaeophyta: Laminariales) on feeding rate and growth in the red abalone *Haliotus rufescens* Swainson. *Journal of Experimental Marine Biology and Ecology, 155,* 263–277.

Wootton, J. T., & Power, M. E. (1993). Productivity, consumers and the structure of a river food chain. *Proceedings of the National Academy of Sciences USA, 90,* 1384–1387.

Worm, B., & Myers, R. A. (2003). Meta-analysis of cod-shrimp interactions reveals top-down control in oceanic food webs. *Ecology, 84,* 162–173.

Zaret, T. M., & Paine, R. T. (1973). Species introduction in a tropical lake. *Science, 182,* 449–455.

Zimov, S., Chuprynin, V. I., Oreshko, A. P., Chapin, S. F., Reynolds, J. F., & Chapin, C. (1995). Steppe-tundra transition: A herbivore-driven biome shift at the end of the Pleistocene. *American Naturalist, 146,* 765–794.

INDEX